Reveal MATH®

Student Edition
Grade 2 • Volume 2

Back cover: Jason Edwards/Getty Images

mheducation.com/prek-12

Copyright © 2022 McGraw Hill

All rights reserved. No part of this publication may be reproduced or distributed in any form or by any means, or stored in a database or retrieval system, without the prior written consent of McGraw Hill, including, but not limited to, network storage or transmission, or broadcast for distance learning.

Send all inquiries to:
McGraw Hill
8787 Orion Place
Columbus, OH 43240

ISBN: 978-0-07-683905-6
MHID: 0-07-683905-2

Printed in the United States of America.

5 6 7 8 9 QSX 24 23 22

Contents in Brief

Volume 1

1. Math Is… 1
2. Place Value to 1,000 31
3. Patterns within Numbers 61
4. Meanings of Addition and Subtraction 99
5. Strategies to Fluently Add within 100 149
6. Strategies to Fluently Subtract within 100 .. 199

Glossary G1

Volume 2

7. Measure and Compare Lengths 1
8. Measurement: Money and Time 55
9. Strategies to Add 3-Digit Numbers 85
10. Strategies to Subtract 3-Digit Numbers 123
11. Data Analysis 169
12. Geometric Shapes and Equal Shares 203

Glossary G1

Welcome to *Reveal Math*!

We are excited that you have made us part of your math journey.

Throughout the school year, you will explore new concepts and develop new skills. You will expand your math thinking and problem-solving skills. You will be encouraged to persevere as you solve problems, working both on your own and with your classmates.

With *Reveal Math*, you will experience activities to spark your curiosity and challenge your thinking. In each lesson, you will engage in sense-making activities that will make you a better problem solver. You will have different learning experiences to help you build understanding.

We look forward to revealing to you the wonder and excitement of math.

The *Reveal Math* authors

The *Reveal Math* Authorship Team

McGraw-Hill teamed up with expert mathematicians to create a program centered around you, the student, to make sure each and every one of you can find joy and understanding in the math classroom.

Ralph Connelly, Ph.D.
Authority on the development of early mathematical understanding.

Annie Fetter
Advocate for student ideas and student thinking that foster strong problem solvers.

Linda Gojak, M.Ed.
Expert in both theory and practice of strong mathematics instruction.

Sharon Griffin, Ph.D.
Champion for number sense and the achievement of all students.

Ruth Harbin Miles, Ed.S.
Leader in developing teachers' math content and strategy knowledge.

Susie Katt, M.Ed.
Advocate for the unique needs of our youngest mathematicians.

Nicki Newton, Ed.D.
Expert in bringing student-focused strategies and workshops into the classroom.

John SanGiovanni, M.Ed.
Leader in understanding the mathematics needs of students and teachers.

Raj Shah, Ph.D.
Expert in both theory and practice of strong mathematics instruction.

Jeff Shih, Ph.D.
Advocate for the importance of student knowledge.

Cheryl Tobey, M.Ed.
Facilitator of strategies that drive informed instructional decisions.

Dinah Zike, M.Ed.
Creator of learning tools that make connections through visual-kinesthetic techniques.

Unit 7

Measure and Compare Lengths

Unit Opener: STEM in Action 1

IGNITE! Which Path Is the Shortest? 2

Lessons

7-1 Measure Length with Inches 3
7-2 Measure Length with Feet and Yards 7
7-3 Compare Lengths Using Customary Units 11
7-4 Relate Inches, Feet, and Yards 15
7-5 Estimate Length Using Customary Units 19
7-6 Measure Length with Centimeters and Meters 23
7-7 Compare Lengths Using Metric Units 27
7-8 Relate Centimeters and Meters 31

Math Probe Relating Measurement 35

7-9 Estimate Length Using Metric Units 37
7-10 Solve Problems Involving Length 41
7-11 Solve More Problems Involving Length 45

Unit Review 49
Fluency Practice 53

Unit 8

Measurement: Money and Time

Unit Opener: STEM in Action 55

IGNITE! How Many Coins? 56

Lessons

   8-1 Understand the Values of Coins 57

   8-2 Solve Money Problems Involving Coins 61

Math Probe Counting Coins 65

   8-3 Solve Money Problems Involving Dollar Bills and Coins 67

   8-4 Tell Time to the Nearest Five Minutes 71

   8-5 Be Precise When Telling Time 75

Unit Review 79

Fluency Practice 83

Unit 9

Strategies to Add 3-Digit Numbers

Unit Opener: STEM in Action 85

IGNITE! Greatest and Least Sums 86

Lessons
- 9-1 Use Mental Math to Add 10 or 100 87
- 9-2 Represent Addition with 3-Digit Numbers 91
- 9-3 Represent Addition with 3-Digit Numbers with Regrouping 95
- 9-4 Decompose Addends to Add 3-Digit Numbers 99
- 9-5 Decompose One Addend to Add 3-Digit Numbers . . . 103
- 9-6 Adjust Addends to Add 3-Digit Numbers 107
- 9-7 Explain Addition Strategies 111

Math Probe Addition Problems 115
Unit Review 117
Fluency Practice 121

Unit 10

Strategies to Subtract 3-Digit Numbers

Unit Opener: STEM in Action 123

IGNITE! Greatest and Least Differences 124

Lessons

   10-1 Use Mental Math to Subtract 10 or 100 125

   10-2 Represent Subtraction with
         3-Digit Numbers 129

   10-3 Decompose One 3-Digit Number to Count Back 133

   10-4 Count On to Subtract 3-Digit Numbers 137

   10-5 Regroup Tens 141

   10-6 Regroup Tens and Hundreds 145

   10-7 Adjust Numbers to Subtract 3-Digit Numbers 149

   10-8 Explain Subtraction Strategies 153

   10-9 Solve Problems Involving Addition
         and Subtraction 157

Math Probe Addition and Subtraction Problems 161

Unit Review 163

Fluency Practice 167

Unit 11

Data Analysis

Unit Opener: STEM in Action 169

IGNITE! Mystery Data 170

Lessons
- 11-1 Understand Picture Graphs 171
- 11-2 Understand Bar Graphs 175
- 11-3 Solve Problems Using Bar Graphs 179
- 11-4 Collect Measurement Data 183
- 11-5 Understand Line Plots 187

Math Probe Reading Line Plots 191
- 11-6 Show Data on a Line Plot 193

Unit Review 197

Fluency Practice 201

Unit 12

Geometric Shapes and Equal Shares

Unit Opener: STEM in Action 203
IGNITE! Prove Me Wrong! 204

Lessons

12-1 Recognize 2-Dimensional Shapes
by Their Attributes 205

12-2 Draw 2-Dimensional Shapes from
Their Attributes 209

12-3 Recognize 3-Dimensional Shapes
by Their Attributes 213

12-4 Understand Equal Shares 217

Math Probe Partitioning Shapes 221

12-5 Relate Equal Shares 223

12-6 Partition a Rectangle into Rows and Columns 227

Unit Review 231
Fluency Practice 235

Jump into Learning!

You can find all the resources you need from your **Student Dashboard**.

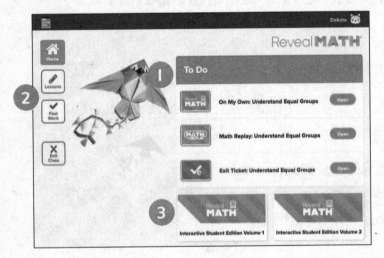

1. See your work in the To-Do List.
2. See the work you already completed.
3. Go to your Interactive Student Edition.

You can use your **Interactive Student Edition** for all your math work.

1. Use the slide numbers to find your page number.
2. Type or draw to work out problems.
3. Check your answers as you go.

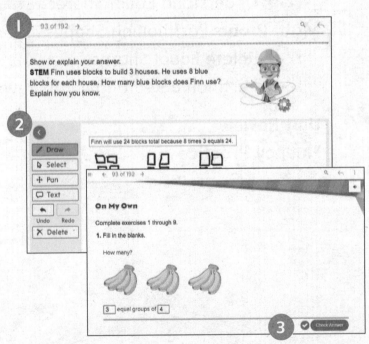

Access Lesson Supports Online!

You can also use these to support you while you practice.

Need an Instant Replay of the Lesson Content?

Each lesson has a **Math Replay** video that provides a 1–2 minute overview of the lesson concept.

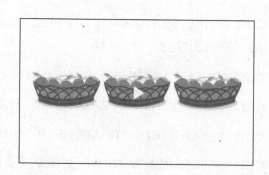

Virtual Tools to Help You Problem Solve

You can access the eToolkit at any time from your Student Dashboard. You can access these tools:

- Counters
- Base-Ten Blocks
- Array Builder
- Fraction Model
- Bucket Balance
- Geometry Sketch
- Money
- Fact Triangles
- Number Line
- and more!

Key Concepts and Learning Objectives

Key Concept **Habits of Mind and Classroom Norms**

- I can make sense of problems and think about numbers and quantities. (Unit 1)
- I can share my thinking with my classmates. (Unit 1)
- I can make sense of problems. (Unit 1)
- I can use patterns to solve problems. (Unit 1)
- I can describe my math story. (Unit 1)
- I can work well with my classmates. (Unit 1)

Key Concept **Addition and Subtraction**

- I can write equations to describe arrays. (Unit 3)
- I can represent and solve one- and two-step word problems using addition and subtraction strategies. (Units 4, 5, 6, 9, 10)
- I can add addends in any order to find the sum. (Unit 5)
- I can add and subtract fluently within 20. (Units 5, 6)
- I can use tools to help me add and subtract. (Units 5, 6)
- I can add and subtract 2-digit and 3-digit numbers with and without regrouping. (Units 5, 6, 9, 10)
- I can mentally add 10 and 100 to a 3-digit number and subtract 10 and 100 from a 3-digit number. (Units 9, 10)
- I can explain how to use strategies to add and subtract 3-digit numbers. (Units 9, 10)

Key Concept Whole Numbers

- I can identify the digits in a 3-digit number. (Unit 2)
- I can read and write numbers to 1,000. (Unit 2)
- I can decompose 3-digit numbers in different ways. (Unit 2)
- I can compare 3-digit numbers. (Unit 2)
- I can identify and describe patterns when counting by 1s, 5s, 10s, and 100s. (Unit 3)
- I can determine the value of a group of coins. (Unit 8)
- I can tell time from analog and digital clocks. (Unit 8)

Key Concept Measurement

- I can measure and compare lengths using customary and metric units. (Unit 7)
- I can use everyday items to help estimate length in customary and metric units. (Unit 7)
- I can solve problems involving length. (Unit 7)
- I can collect measurement data. (Unit 11)
- I can interpret data on a line plot. (Unit 11)
- I can make a line plot to show data. (Unit 11)

Key Concept Describe and Analyze Shapes

- I can describe 2-dimensional and 3-dimensional shapes. (Unit 12)
- I can identify equal shares. (Unit 12)
- I can partition 2-dimensional shapes into equal shares. (Unit 12)
- I can partition rectangles into rows and columns of equal-sized squares. (Unit 12)

Math is...

How would you complete this sentence?

Math is...

Math is not just adding and subtracting.

Math is...
- working together
- finding patterns
- sharing ideas
- listening thoughtfully to our classmates
- sticking with a task even when it is a little challenging

In *Reveal Math*, you will develop the habits of mind that strong doers of math have. You will see that math is all around us.

Let's be Doers of Mathematics

Remember, math is more than getting the right answer. It is a tool for understanding the world around you. It is a language to communicate and collaborate. Be mindful of these prompts throughout the year to access the power of math.

1. **Math is...** Mine
   - Mindset

2. **Math is...** Exploring and Thinking
   - Planning
   - Connections
   - Thinking

3. **Math is...** My World
   - In My World
   - Modeling
   - Choosing Tools

4. **Math is...** Explaining and Sharing
   - Explaining
   - Sharing
   - Precision

5. **Math is...** Finding Patterns
   - Patterns
   - Generalizations

6. **Math is...** Ours
   - Mindset

Math is... Mindset

What can you do to be an active listener?

Explore the Exciting World of STEM!

Ever wonder how math applies in the real world? In every unit, you will learn about a STEM career, from protecting our parks to exploring outer space. You will learn about the STEM career through digital simulations and projects.

STEM Career Kid: Meet Sienna
Let the STEM Career Kid introduce his or her career and talk about the different job responsibilities.

Math In Action: Nutritionist
Watch the Math in Action to see how the math you are learning applies to the real world.

Hi, I'm Sienna.
I want to be a nutritionist to help people eat to feel great!

Unit 7

Measure and Compare Lengths

Focus Question

How can I estimate and measure length in standard units?

Hi, I'm Jordan.

I want to be an animal trainer. I need to measure animals to see if they have grown. I need to understand how to measure and which tools are appropriate for the job.

STEM video | GO ONLINE

IGNITE!

Name

Which Path Is the Shortest?

Which path is the shortest?

Use string or a straightedge to measure each path.

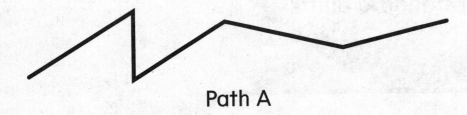

Path A

Path B

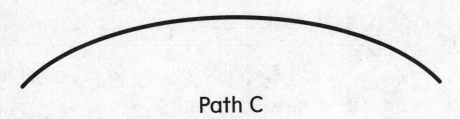

Path C

Straightedge

2 Ignite! • Which Path Is the Shortest?

Lesson 7-1
Measure Length with Inches

Be Curious

How are they the same?
How are they different?

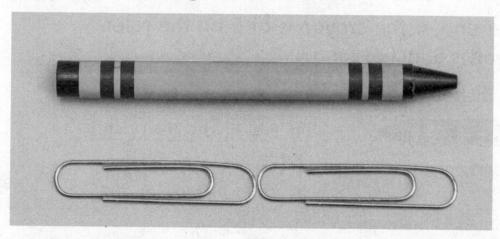

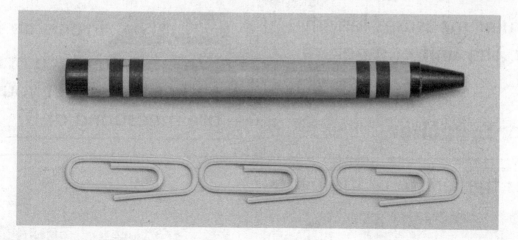

Math is... Mindset
How do you show you understand how others are feeling?

Learn

How can you measure the crayon?

A ruler is a tool to measure length.

Line one end of the crayon with the 0 on the ruler.

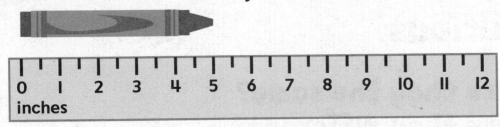

The other end of the crayon is at 5 on the ruler. The crayon is 5 inches long.

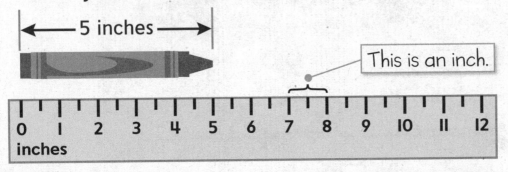

This is an inch.

An inch ruler measures length in **inches**. The **unit** of measure is inches.

Math is... Precision

Why do you place one end of the object you are measuring at 0?

Work Together

What is the length of the pencil?

_____ inches

4 Lesson 1 • Measure Length with Inches

On My Own

Name _____

What is the length of the object? Use an inch ruler to measure.

1.

   _____ _____

2.

   _____ _____

3.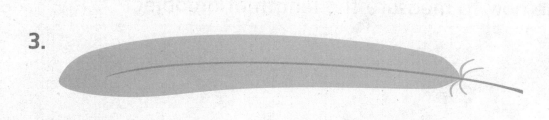

   _____ _____

4. **Will this glue stick fit into a box that has a length of 3 inches? Explain.**

5. **Error Analysis** Gina says the stapler is 11 inches long. Is she correct? If not, how can she find the correct length?

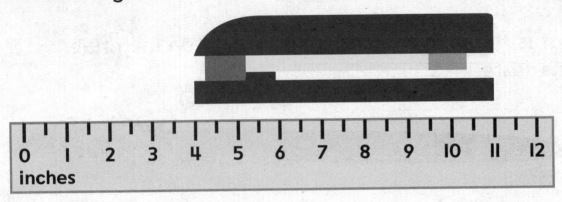

6. **Extend Your Thinking** How would you explain to someone how to measure the length of an object in inches?

 **Reflect**

When are inches a good unit to use when measuring the length of an object?

Math is... Mindset

How did you show you understand how others are feeling?

Lesson 7-3
Compare Lengths Using Customary Units

Be Curious

What question could you ask?

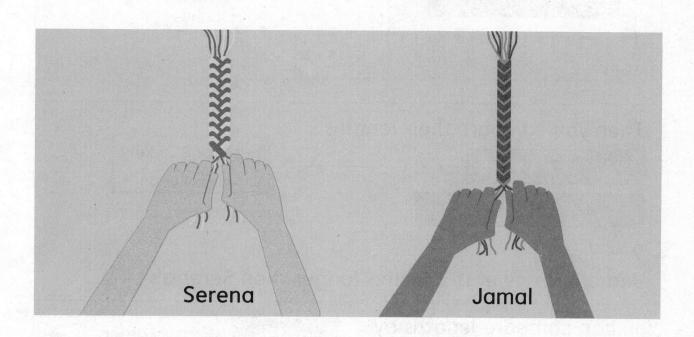

Math is... Mindset

What helps you stay focused on your work?

Learn

Serena thinks the two bracelets are the same length. Jamal thinks his bracelet is longer.

How can you find out who is correct?

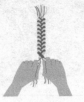 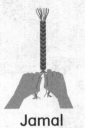

Serena   Jamal

You can measure the lengths of the bracelets.

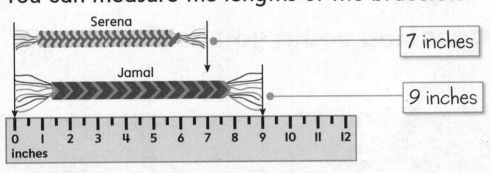

Then you compare their lengths.

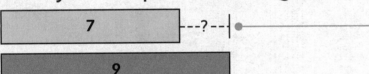

Subtract to find the difference.

$9 - 7 = 2$

Jamal's bracelet is 2 inches longer than Serena's.

You can compare lengths by subtracting the measurements to find the difference.

Math is... Explaining

Would using a different unit of measure result in the same comparison? Explain.

 Work Together

Mrs. Green's desk is 6 feet long. Her bulletin board is 11 feet long. How can you compare the two lengths?

12 Lesson 3 • Compare Lengths Using Customary Units

On My Own

Name _____

How can you compare the lengths? Write an equation to compare the lengths.

1. Fred jumps 3 feet. Jeff jumps 6 feet.

   ____ − ____ = ____

2. Hadia's kitchen is 4 yards long. Her family room is 9 yards long.

   ____ − ____ = ____

Which object is longer? Write an equation and the answer.

3.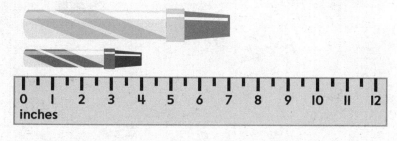

   ____ − ____ = ____

   The top marker is ____ inches longer than the bottom marker.

4.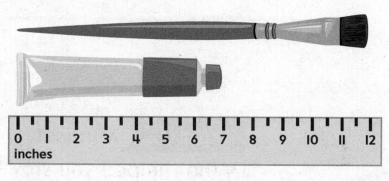

   ____ − ____ = ____

   The paintbrush is ____ inches longer than the paint.

Unit 7 • Measure and Compare Lengths   13

5. **Error Analysis** The length of Gary's swimming pool is 14 feet. The length of Paul's swimming pool is 18 feet. Paul thinks his pool is 32 feet longer than Gary's pool because 18 + 14 = 32. How do you respond to Paul?

6. **Extend Your Thinking** Write a word problem that involves comparing the lengths of two objects that are measured in feet. Then solve your problem.

How can you find the difference in length between two objects?

Math is... Mindset

What helped you stay focused on your work?

Lesson 7-4
Relate Inches, Feet, and Yards

Be Curious

**What do you notice?
What do you wonder?**

I notice

Math is... Mindset
How can working as a team help you achieve your goal?

Learn

Emilia measures a whiteboard to be 6 feet long.

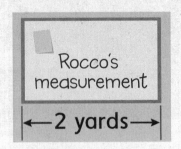
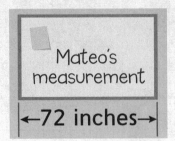

How does her measurement relate to Rocco's and Mateo's measurements?

You can measure length in inches, feet, or yards.

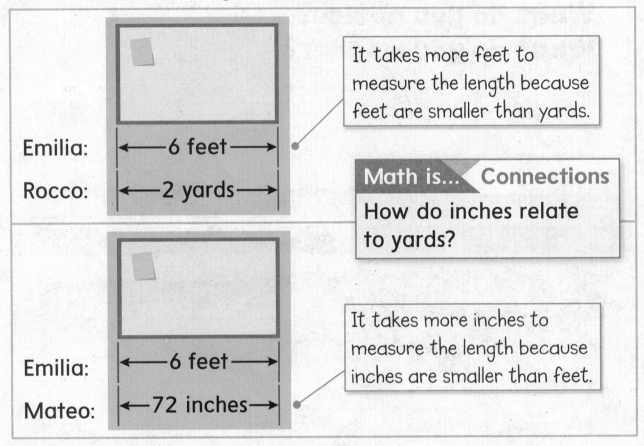

It takes more feet to measure the length because feet are smaller than yards.

Math is... Connections

How do inches relate to yards?

It takes more inches to measure the length because inches are smaller than feet.

The smaller the unit, the more units are needed to measure an object's length.

 Work Together

Measure the length of your desk in inches and in yards. Which unit is smaller? How do inches relate to yards?

16 Lesson 4 • Relate Inches, Feet, and Yards

On My Own

Name _____

1. What is the length of the classroom wall in yards?

   _____ _____

   Will the measurement of the classroom wall have fewer feet or fewer yards? Circle the answer.

   feet   yards

2. What is the length of the bookshelf in inches?

   _____ _____

   Will the measurement of the bookshelf have more inches or more yards? Circle the answer.

   inches   yards

3. What is the length of the whiteboard in feet?

   _____ _____

   Will the measurement of the whiteboard have fewer inches or fewer feet? Circle the answer.

   inches   feet

4. **Error Analysis** Roshni and Shingi want to measure the trumpet using inches and feet. Roshni thinks there will be more feet. Shingi thinks there will be more inches. How do you respond to them?

5. **Extend Your Thinking** Natalie measures the length of her garden in feet. Then she measures it in yards. Are there more feet or yards? Explain your thinking.

 **Reflect**

How are inches, feet, and yards related?

Math is... Mindset

How has working as a team helped you achieve your goal?

Lesson 7-5
Estimate Length Using Customary Units

Be Curious

**What do you notice?
What do you wonder?**

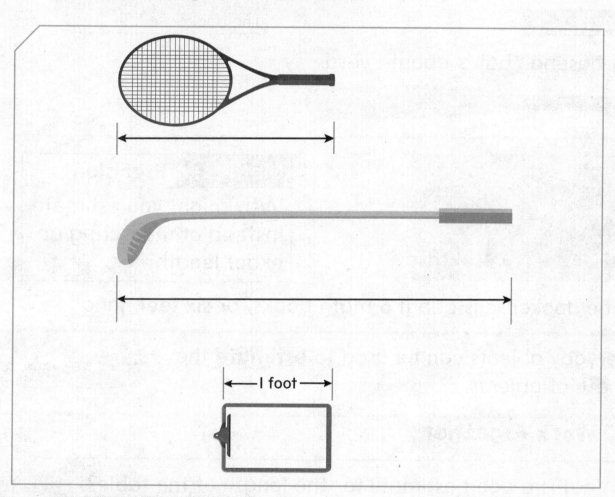

Math is... Mindset

What makes you feel frustrated in math?

Unit 7 • Measure and Compare Lengths 19

Learn

Bryce wants to find the length of the bookcase. He does not have a measuring tool.

How can Bryce find the length of the bookcase?

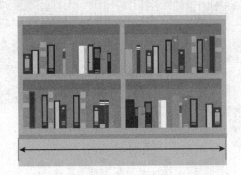

A paper clip is about 1 inch.

A math book is about 1 foot.

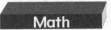

A baseball bat is about 1 yard.

> Which object can help you find the length of the bookcase?

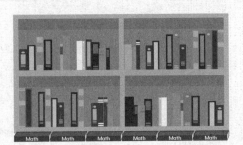

Math is... Precision

Why might you estimate instead of measuring an exact length?

The bookcase is about 6 math books, or six feet, long.

Everyday objects can be used to **estimate** the length of objects.

Work Together

What is a good estimate for the length of the table?

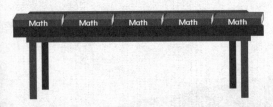

_____ feet

On My Own

Name _____

Which everyday item can you use to estimate the length of the object? Circle the answer.

1. **marker**

   paper clip   math book

2. **door (top to bottom)**

   color tile   science book

3. **whiteboard**

   paper clip   clipboard

4. **area rug**

   color tile   math book

5. **sticky notepad**

   color tile   science book

6. **bracelet**

   paper clip   clipboard

---

7. About how long is the glue bottle? Estimate the length.

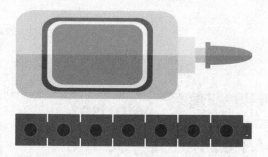

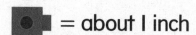 = about 1 inch

_____

8. About how long is the wire? Estimate the length.

 = about 1 foot

_____

9. **Error Analysis** Mae uses a paper clip to estimate the length of her hairbrush. She says her hairbrush is about 10 feet long. Is Mae's estimate reasonable? Explain why or why not.

10. **Extend Your Thinking** Tom's dad walked heel to toe from one side of their family room to the other. What do you think he was trying to do?

Reflect

How can you use everyday items to estimate length in inches and feet?

Math is... Mindset
What made you feel frustrated in math today?

Lesson 7-6
Measure Length with Centimeters and Meters

Be Curious

How are they the same?
How are they different?

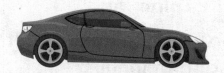

Math is... Mindset

How do you help build a productive classroom culture?

Learn

How can you measure the pen and the bat?

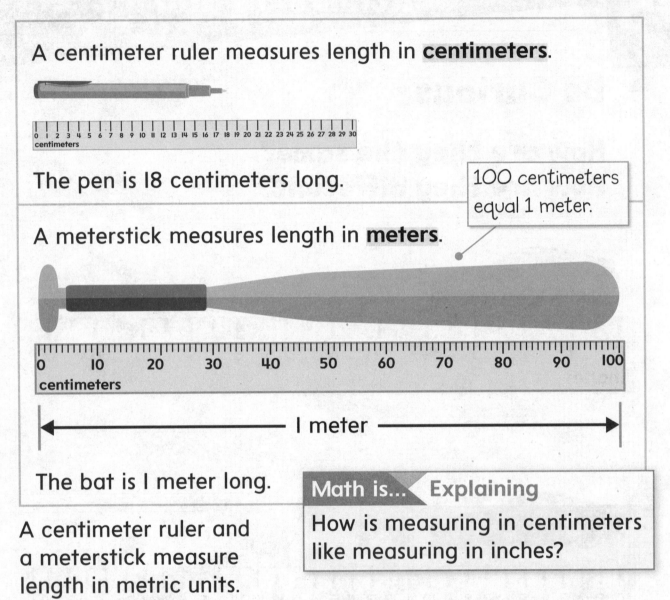

A centimeter ruler measures length in **centimeters**.

The pen is 18 centimeters long.

100 centimeters equal 1 meter.

A meterstick measures length in **meters**.

1 meter

The bat is 1 meter long.

A centimeter ruler and a meterstick measure length in metric units.

Math is... Explaining

How is measuring in centimeters like measuring in inches?

Work Together

What is the length of your desk in centimeters?

 centimeters

24   Lesson 6 • Measure Length with Centimeters and Meters

On My Own

Name _____

What is the length of the object in centimeters? Use a centimeter ruler to measure.

1.

_____ _____

2. ⌬

_____ _____

3.

_____ _____

Unit 7 • Measure and Compare Lengths   25

What is the length of the object in meters? Use a meterstick to measure.

4. classroom wall

5. bookshelf

_____ _____   _____ _____

6. **STEM Connection** Jordan is volunteering with an animal trainer at the zoo. His job is to measure the length of the penguin habitat. Should he use a centimeter ruler or a meterstick to measure? Explain.

7. **Extend Your Thinking** Use a centimeter ruler to draw a pencil that is 10 centimeters long.

Reflect

What do you know about measuring length in centimeters and meters?

Math is... Mindset

How did you help build a productive classroom culture?

Lesson 7-7
Compare Lengths Using Metric Units

Be Curious

What could the question be?

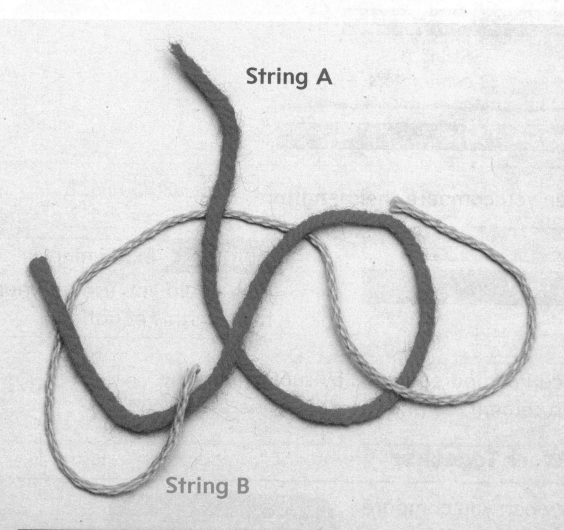

String A

String B

Math is... Mindset
What do you do to stay focused on your work?

Unit 7 • Measure and Compare Lengths

Learn

Angela has two strings. She uses the longer string for her project.

How much longer is the string she uses?

You can measure the lengths of the strings.

String A
|------- 19 centimeters -------|

String B
|----------- 23 centimeters -----------|

Then you compare their lengths.

| 19 | --?--|
| 23 |

Subtract to find the difference.

$23 - 19 = 4$

Math is... Explaining

How could you use addition to solve this equation?

You can compare lengths by subtracting the measurements to find the difference.

Work Together

How can you compare the two lengths?

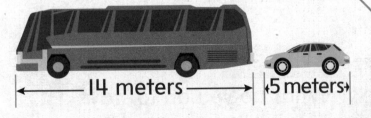

14 meters • 5 meters

28 Lesson 7 • Compare Lengths Using Metric Units

On My Own

Name _____

How can you compare the lengths? Write the equation.

1. Mary's driveway is 19 meters long. John's driveway is 27 meters long.

   ____ − ____ = ____

2. Danielle's scarf is 59 centimeters long. Corey's scarf is 71 centimeters long.

   ____ − ____ = ____

3. How much longer is the spoon than the fork?

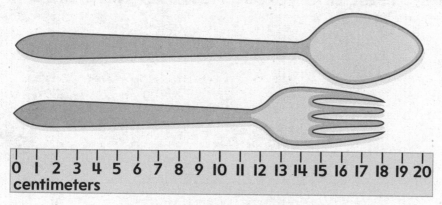

   _____

4. How much shorter is the school bus than the train car?

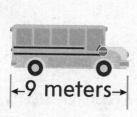 

   ←9 meters→   ←18 meters→

   _____

Unit 7 • Measure and Compare Lengths 29

5. **STEM Connection** The length of Deven's computer cord is 4 meters. The length of his speaker wire is 11 meters. How much shorter is the computer cord than the speaker wire?

_____ _____

6. A paper clip is 4 centimeters long. A tube of lip balm is 6 centimeters long. How much longer is the tube of lip balm than the paper clip?

_____ _____

7. **Extend Your Thinking** How is comparing objects in centimeters and meters the same as comparing objects in inches, feet, and yards? How is it different?

Reflect

How do you know if one object is longer or shorter than another object?

Math is... Mindset
What helped you stay focused on your work today?

Lesson 7-8
Relate Centimeters and Meters

Be Curious

Tell me everything you can.

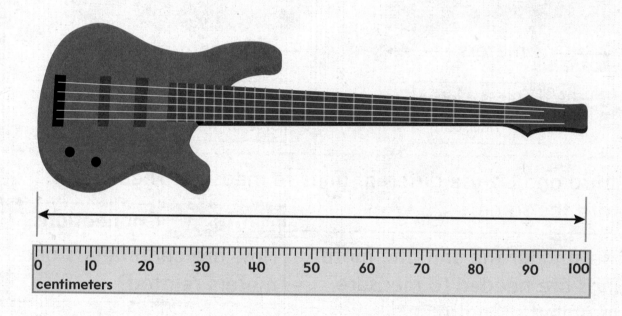

| Math is... | Mindset |

What can you do to show respect for your classmates?

Unit 7 • Measure and Compare Lengths

Learn

Hiro and Liz measure the same bulletin board. Hiro says the length is 2 meters. Liz says the length is 200 centimeters.

How can you respond to Hiro and Liz?

Hiro uses a meterstick to measure.

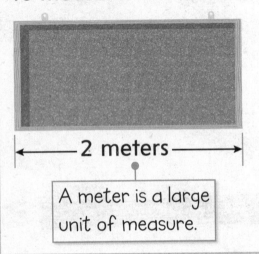

← 2 meters →

A meter is a large unit of measure.

Liz uses a centimeter ruler to measure.

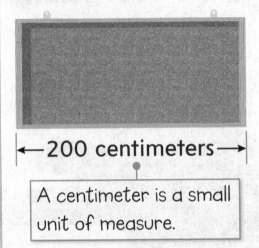

← 200 centimeters →

A centimeter is a small unit of measure.

Hiro and Liz use different units to measure. The lengths are the same.

The smaller the unit, the more units are needed to measure an object's length.

Math is... Connections

How are centimeters and meters related?

Work Together

Measure the length of an object in the classroom that you think is about 1 meter long. Then measure it in centimeters. Which unit is smaller? How do meters relate to centimeters?

On My Own

Name _____

1. What is the length of the whiteboard in meters?

   _____

   Will the measurement of the whiteboard have more centimeters or more meters? Circle the answer.

   centimeters   meters

2. What is the length of the teacher's desk in centimeters?

   _____

   Will the measurement of the desk have fewer centimeters or fewer meters? Circle the answer.

   centimeters   meters

3. The length of a picnic table is measured in meters and centimeters. Will the measurement have more meters or more centimeters? Circle the answer.

   meters   centimeters

4. Hideki measured the length of his car in centimeters. Then he measured it in meters. Are there more centimeters or more meters? Explain your thinking.

Unit 7 • Measure and Compare Lengths

5. Rae measures the length of her bed in centimeters. Then she measures it in meters. Are there fewer centimeters or fewer meters? Explain your thinking.

6. **Error Analysis** Khal and his sister want to measure Khal's bike. Khal thinks there will be more meters. His sister thinks there will be more centimeters. How do you respond to them?

7. **Extend Your Thinking** Will there always be more centimeters than meters in two measurements of the same object? How do you know?

Reflect

What is the relationship between centimeters and meters?

Math is... Mindset

How did you show respect for your classmates today?

Unit 7
Relating Measurement

Name _____

Determine the unit used to measure the length or height of an object.

1. Two students measured the **length of a laptop**.
   Student 1 says: *I got 1.*
   Student 2 says: *I got 12.*
   Who likely measured using **inches** as the unit?
   Circle the answer.
   a. Student 1
   b. Student 2
   c. Neither student
   d. Both students

   Explain your choice.

2. Two students measured the **height of a bookcase**.
   Student 1 says: *I got 3.*
   Student 2 says: *I got 36.*
   Who likely measured using **feet** as the unit?
   Circle the answer.
   a. Student 1
   b. Student 2
   c. Neither student
   d. Both students

   Explain your choice.

3. Two students measured the **height of a table**.

   Student 1 says: *I got 1.*

   Student 2 says: *I got 100.*

   Who likely measured using **meters** as the unit?

   Circle the answer.
   a. Student 1
   b. Student 2
   c. Neither student
   d. Both students

   Explain your choice.

4. Two students measured the **length of a notebook**.

   Student 1 says: *I got 1.*

   Student 2 says: *I got 30.*

   Who likely measured using **centimeters** as the unit?

   Circle the answer.
   a. Student 1
   b. Student 2
   c. Neither student
   d. Both students

   Explain your choice.

Reflect On Your Learning

Lesson 7-9
Estimate Length Using Metric Units

Be Curious

**What do you notice?
What do you wonder?**

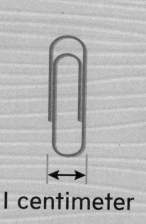

I centimeter

Math is... Mindset
How can you be flexible in your thinking?

Unit 7 • Measure and Compare Lengths

Learn

How can Erin estimate the length of a classroom wall?

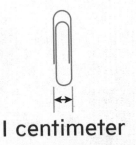
I centimeter

← I meter →

You can use your arm span to estimate the length of the wall.

Math is... Generalizations

When might estimating length be useful?

The wall is about 7 meters long.

You can use everyday objects to help you estimate length in centimeters and meters.

 Work Together

A staple is about I centimeter long. What is a good estimate for the length of the glue stick?

_____ centimeters

38 Lesson 9 • Estimate Length Using Metric Units

On My Own

Name _____

Which everyday item can you use to estimate the length of the object? Circle the answer.

1. **earring**

    width of paper clip   arm span

2. **house**

    staple   baseball bat

3. **lip balm**

    unit cube   baseball bat

4. **remote control**

    unit cube   arm span

Which unit would you use to measure the length of the object? Circle the answer.

5. **cell phone**

    centimeter   meter

6. **truck**

    centimeter   meter

7. **vegetable garden**

    centimeter   meter

8. **bar of soap**

    centimeter   meter

9. The arm span of a second grader is about 1 meter long. About how long is the swing set? Estimate the length.

10. **Extend Your Thinking** Would you use estimated lengths to build a bookcase? Why or why not? Explain your thinking.

Reflect

How can you use everyday items to estimate length in centimeters and meters?

Math is... Mindset

How were you flexible in your thinking today?

Lesson 7-10
Solve Problems Involving Length

Be Curious

What is the question?

The art teacher has some red ribbon.
He also has some yellow ribbon.

Math is... Mindset
What is your goal for today?

Learn

The art teacher has 18 feet of red ribbon.
He has 13 feet of yellow ribbon.

How much ribbon does the art teacher have in all?

You can make a drawing to represent the problem.

← total length of ribbon →
| 18 feet | 13 feet |

Write an equation to match the drawing. 18 + 13 = ? 31

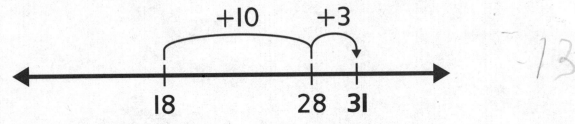

−13

The art teacher has **31 feet** of ribbon.

You can solve addition and subtraction word problems involving length.

Math is... Connections
What are some strategies you can use to add lengths?

 Work Together

Adele has 33 yards of ribbon. She uses some ribbon. Now she has 16 yards of ribbon. Make a drawing and write an equation to find how much ribbon Adele uses.

42 Lesson 10 • Solve Problems Involving Length

On My Own

Name _____

**Which equation represents the problem?
Circle the answer.**

1. The scout leader has 16 feet of brown rope. She has 15 feet of yellow rope. How much rope does the scout leader have in all?

   16 + 15 = ? 31 ✓   16 − 15 = ? 1 ✗

2. The length of Matt's desk is 42 inches. The length of Denise's desk is 6 inches shorter than Matt's desk. What is the length of Denise's desk?

   6 + 42 = ? ✗   42 − 6 = ? 36 ✓

3. **STEM Connection** Erik is playing a video game. He has to move 36 meters to win. In Round 1, He moves 19 meters. How much more does he need to move to win? Represent and solve the problem with a drawing and an equation.

Unit 7 • Measure and Compare Lengths   43

4. Mr. Jones has some paper for the bulletin board. Then he finds 8 more feet of paper. Now, he has 20 feet of paper. How many feet of paper did Mr. Jones have at first? Represent and solve the problem with a drawing and an equation.

5. Bea has 45 yards of fabric. She uses some of the fabric. Now she has 18 yards of fabric left. Explain how you can find how much fabric Bea used.

6. **Extend Your Thinking** Karen's bedroom is 16 feet long. Tom's bedroom is 5 feet longer than Karen's bedroom. What is the length of both bedrooms combined? Explain your thinking.

Reflect

How can making a drawing help you solve addition and subtraction word problems involving length?

Math is... Mindset

What helped you reach your goal for today?

Lesson 7-11
Solve More Problems Involving Length

Be Curious

What question could you ask?

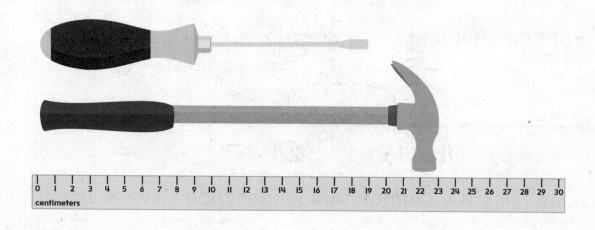

Math is... Mindset

How can you be part of the classroom community?

Unit 7 • Measure and Compare Lengths 45

Learn

Diane draws a line 26 centimeters long.

Oliver draws a line 15 centimeters long.

How much longer is Diane's line than Oliver's line?

You can make a drawing to represent the problem.

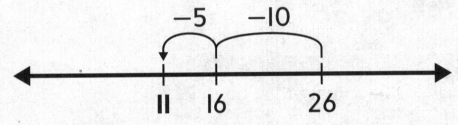

Math is... Connections

What are some strategies to subtract?

Write an equation. $26 - 15 = ?$

Diane's line is **11 centimeters** longer than Oliver's line.

You can solve addition and subtraction word problems involving length.

Work Together

Ethan runs 24 meters. Then he runs some more. Ethan runs a total of 41 meters. Make a drawing and write an equation to find how many more meters Ethan ran.

46   Lesson 11 • Solve More Problems Involving Length

On My Own

Name _____

1. The length of a camper is 33 feet. The length of a pickup truck is 15 feet. How much longer is the camper? Circle the equation you can use to solve the problem.

    $33 - 15 = ?$ $33 + 15 = ?$

2. Cliff sprints 22 meters. Then he sprints some more. In all, he sprints 40 meters. Explain how you can find how many more meters Cliff sprinted.

Make a drawing and write an equation to represent the problem. Use the number line to solve.

3. Val has a piece of yarn 28 inches long. Ty has a piece of yarn 13 inches long. How much longer is Val's piece of yarn than Ty's?

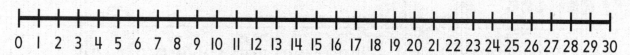

4. A board is 20 centimeters long. Some of the board is cut off and 7 centimeters remain. How much of the board was cut off?

Unit 7 • Measure and Compare Lengths   47

5. **Error Analysis** Tami and Kee solve this problem.

   > Alex builds two wooden trains. The red train is 11 inches long. The green train is 4 inches long. How much shorter is the green train?

   Tami writes: $11 - 4 = ?$; 7 inches shorter.

   Kee writes: $4 + ? = 11$; 7 inches shorter.

   Who is correct? Explain.

6. **Extend Your Thinking** Write an addition word problem that involves length, for which the first addend is unknown. Then solve the problem.

Reflect

How can using a number line help you solve problems involving length?

Math is... Mindset

How were you part of the classroom community today?

Unit Review   Name _____

Vocabulary Review

Use the vocabulary to complete each sentence.

centimeter   estimate
foot   inches
meter   unit

1. The length of a baseball bat is about one
   _____. (Lesson 7-6)

2. The width of a paper clip is about one
   _____. (Lesson 7-6)

3. A ruler has 12 _____. (Lesson 7-1)

4. 12 inches is the same length as 1 _____.
   (Lesson 7-2)

5. Inches are a _____ of measure. (Lesson 7-1)

6. To find a number close to an exact amount means to
   _____. (Lesson 7-5)

Review

7. A driveway is measured in centimeters and meters. Will the measurements have fewer centimeters or fewer meters? Circle the answer. (Lesson 7-8)

    centimeters   meters

8. What is the length of the feather in inches? (Lesson 7-1)

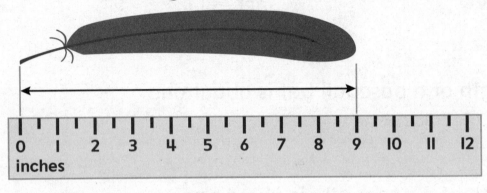

    _____ inches

9. Which tool is best used to measure the length of a bus? Choose the best answer. (Lesson 7-2)

    A. inch ruler   B. centimeter ruler

    C. yardstick   D. tape measure

10. Ana and Trent are drawing pictures with chalk on the driveway. Ana's picture is 63 inches long. Trent's picture is 49 inches long. How much longer is Ana's picture than Trent's picture? (Lesson 7-10)

    _____ inches

11. What is the length of the paintbrush in centimeters?
(Lesson 7-6)

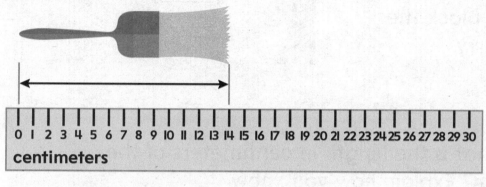

_____ centimeters

12. Frida runs 47 meters and Diego runs 83 meters. How many fewer meters does Frida run than Diego?
(Lesson 7-10)

_____ meters

13. Jeri digs a ditch that is 8 yards long. Lynn digs a ditch that is 5 yards long. What is the difference in lengths?
(Lesson 7-3)

_____ yards

14. Which item can be used to estimate the length of a pair of scissors? Choose the correct answer. (Lesson 7-9)

A.

B.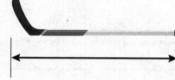

C.

D.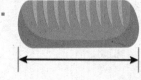

Performance Task

An animal trainer is measuring the length of a dog's mouth and the length of a block the dog can carry.

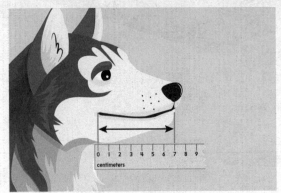

Part A: What is the length in centimeters of the dog's mouth? Explain how you know.

Part B: What is a good estimate for the length in centimeters of the block? Explain how you know.

Part C: How much longer is the dog's mouth than the block? Explain your answer.

How do you measure length?

Unit 7
Fluency Practice

Name _____

Fluency Strategy

> You can use a doubles fact to help you find a difference.
>
> 11 − 5 = ?
>
> Think: I know 5 + 5 = 10.
>
> 11 is 1 more than 10.
>
> Add 1 to one of the addends in the doubles fact: 5 + 1 = 6
>
> So, 11 − 5 = 6.

1. What doubles fact helps you subtract 17 − 8? Find the difference. Explain how you found the difference.

Fluency Flash

2. How can you use a doubles fact to subtract? Write the numbers.

   14 − 8 = ?

   Doubles fact: 8 + _____ = _____

   14 is 2 less than _____.

   Subtract _____ from one of the addends in the doubles fact: 8 − _____ = _____.

   So, 14 − 8 = _____.

Fluency Check

What is the sum or difference?

3. $12 - 5 =$ _____

4. $14 - 9 =$ _____

5. $5 + 6 =$ _____

6. $16 - 7 =$ _____

7. $15 - 8 =$ _____

8. $11 - 6 =$ _____

9. $8 + 9 =$ _____

10. $16 - 9 =$ _____

11. $14 - 6 =$ _____

12. $7 + 8 =$ _____

13. $17 - 8 =$ _____

14. $4 + 5 =$ _____

Fluency Talk

How can you use a doubles fact to subtract $13 - 6$? Explain your thinking.

How is using a doubles fact to subtract like using a doubles fact to add? How is it different? Explain.

Unit 8

Measurement: Money and Time

Focus Question

How can I measure with money and time?

Hi, I'm C.J.

I want to be a statistician. As an experiment, I ask everyone in my family how much change they have in their pockets. They have 10 pennies, 12 nickels, and 5 quarters.

Name _____

How Many Coins?

Use as few coins as possible to make the amount.

Amount	Dimes	Nickels	Pennies	Total Number of Coins
1 cent				
2 cents				
3 cents				
4 cents				
5 cents				
6 cents				
7 cents				
8 cents				
9 cents				
10 cents				
11 cents				
12 cents				
13 cents				
14 cents				
15 cents				
16 cents				
17 cents				
18 cents				
19 cents				
20 cents				

Lesson 8-3
Solve Money Problems Involving Dollar Bills and Coins

Be Curious

Which doesn't belong?

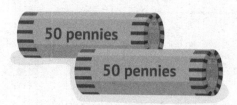

Math is... Mindset

How do you show that you value the ideas of others?

Learn

Ava has $41 in her piggy bank. She adds $29.

What are some ways to show the amount of money Ava has?

You can use **dollar bills**.

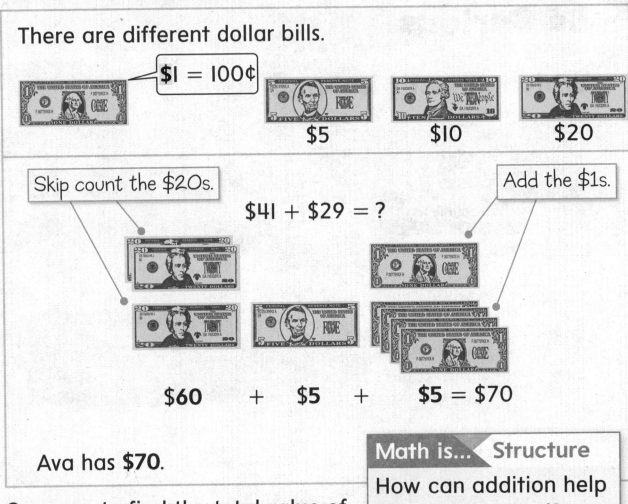

Ava has **$70**.

One way to find the total value of dollar bills is to add the values.

Math is... Structure

How can addition help you find the total?

Work Together

Eugene has 67¢. He gives 31¢ to Emma. How much money does Eugene have now?

On My Own

Name _____

What is the value of the group of coins or dollar bills?

1.

   _____ ¢

2.

   $ _____

3.

   $ _____

---

4. Pam has $38 in dollar bills. What dollar bills could she have?

5. David has two $20 bills, one $10 bill, and three $1 bills in his wallet. How much money does he have in all?

6. How can $45 be shown with the fewest number of dollar bills? Explain.

7. **Error Analysis** Drew solves this problem.

> Joan has $16 in bills. Her mom gives her $15 more in bills. How much money does Joan have now? What dollar bills could she have?

Drew thinks Joan has $31. He says Joan could have one $20 bill, one $10 bill, and one $1 bill. How do you respond to Drew? Explain.

8. **Extend Your Thinking** Kelly has three $1 bills, two quarters, two dimes, and two nickels. She wants to buy a purse that costs $4. Does she have enough money? Explain.

Reflect

How can you find the total value of a group of mixed dollar bills or coins?

Math is... Mindset

How did you show that you value the ideas of others?

Lesson 8-4
Tell Time to the Nearest Five Minutes

Be Curious

Tell me everything you can.

Math is... Mindset
How do you help make everyone feel safe in class?

Learn

The clocks show what time Rosita begins school and eats lunch each school day.

What time does Rosita start school and eat lunch at school?

School starts   Lunch time

Analog and **digital clocks** are used to tell time.

Count by 5s to 5 to find how many minutes past 8 o'clock.

The minute hand is on 5.

Rosita begins school at 8:25.

Count by 5s to 10 to find how many minutes past 11 o'clock.

The minute hand is on 10.

Rosita has lunch at 11:50.

You can tell time to the nearest five minutes on an analog clock by skip counting by 5s.

Math is... Making Sense

What helps you know which is the hour hand?

 Work Together

Zion rides his bike in the morning. Then he plays basketball. Write the time of each activity on the digital clock.

 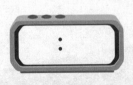    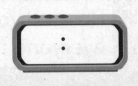

72 Lesson 4 • Tell Time to the Nearest Five Minutes

On My Own

Name ....

1. What time is shown? Circle the digital clock that matches.

What time is shown? Write the time.

2.

3.

____ : ____   ____ : ____

4.

5.

____ : ____   ____ : ____

Unit 8 • Measurement: Money and Time

6. What time is shown? Choose all the correct answers.

A. a quarter to 3:00

B. half past 3:00

C.

D. 6:15

7. **STEM Connection** Erik works on a design for a new video game. He starts working at 7:45. What is another way of writing this time?

8. **Extend Your Thinking** Write and draw the time 6:15 three different ways.

Reflect

How can you tell time to the nearest five minutes?

Math is... Mindset

How did you help make everyone feel safe in class?

Lesson 8-5
Be Precise When Telling Time

Be Curious

What do you notice?
What do you wonder?

| Math is... | Mindset |

What are some ways to build positive relationships with classmates?

Unit 8 • Measurement: Money and Time

Learn

What might you do at 7:00 a.m.?
What might you do at 7:00 p.m.?

You can use a timeline to show the order of events during the day.

wake up · brush teeth · eat lunch · brush teeth · go to sleep

a.m. | p.m.

12:00 a.m. midnight — 12:00 p.m. noon — 12:00 a.m. midnight

You might wake up at 7:00 a.m.

You might brush your teeth at 7:00 p.m.

Math is... Thinking

Why is it important to know if it is a.m. or p.m.?

The time from midnight to noon is represented using **a.m.**

The time from noon to midnight is represented using **p.m.**

Work Together

What time of day is Evan in math class? Write a.m. or p.m. Then explain how you know the time of day.

11:20 ____

76 Lesson 5 • Be Precise When Telling Time

On My Own

Name _____

What time of day does the event take place? Write a.m. or p.m.

1.
3:00 _____

2.
4:30 _____

3.
10:45 _____

4.
9:00 _____

5.
9:30 _____

6.
11:15 _____

7. **Error Analysis** Marissa solves this problem by writing the time on the digital clock.

Patrick is cooking dinner. What time is it?

Has Marissa written the correct time? If not, how could you help her understand the time?

8. **Extend Your Thinking** What event can take place in the a.m.? What event can take place in the p.m.? Label both events on the timeline.

 **Reflect**

Why can time be a.m. or p.m.?

Math is... Mindset

What helped you build positive relationships with classmates?

Unit Review   Name _____

Vocabulary Review

Match each term to the correct coin or dollar bill.

1. dime _____   A.
   (Lesson 8-1)

2. dollar bill _____   B.
   (Lesson 8-3)

3. nickel _____   C.
   (Lesson 8-1)

4. penny _____   D.
   (Lesson 8-1)

5. quarter _____   E.
   (Lesson 8-1)

Unit 8 • Measurement: Money and Time   79

Review

6. Sharif gets these coins as change. How much change does he get? (Lesson 8-2)

7. Elizabeth has 15 nickels. How much money does she have? (Lesson 8-1)

   A. 15 cents

   B. 75 cents

   C. 15 dollars

   D. 75 dollars

8. What time might you play at the park? (Lesson 8-5)

   A. 4:30 p.m.

   B. 2:00 a.m.

   C. 11:15 p.m.

   D. 12:45 a.m.

9. What time does the clock show? (Lesson 8-4)

10. Which activity might happen at 7:30 a.m.? Choose all the correct answers. (Lesson 8-5)

    A. Eat breakfast
    B. Eat dinner
    C. Get on the bus to go to school
    D. Play basketball after school.

11. Bob has 45¢. He buys a piece of gum for 15¢. How much money does Bob have now? (Lesson 8-3)

Performance Task

A statistician wants to buy a book of sports records. The book costs $32.

Part A: The statistician has one $10 bill, three $5 bills, and four $1 bills. How much more money does she need to buy the book?

Part B: The statistician earns enough money to buy the book and a bookmark. She gets 63¢ back in change. What coins could she have gotten back? Show two different ways. Explain your answer.

 **Reflect**

How did you measure with money and time?

Unit 8

Fluency Practice

Name _____

Fluency Strategy

> You can use facts you know to help you find a sum.
> $7 + 4 = ?$
>
> ▶ **One Way:** I can make a 10: $7 + 3 = 10$. Then add 1 more.
> So, $7 + 4 = 11$.
>
> ▶ **Another Way:** I can use the doubles fact: $4 + 4 = 8$.
> 7 is **3 more** than 4. Add $8 + 3$.
> So, $7 + 4 = 11$.

1. How can you use a known fact to find $3 + 5$? What is the sum?

Fluency Flash

How can you use a fact you know to add?
Write the numbers.

2. $6 + 3 = ?$
   I can make a 10.
   $6 + \underline{} = \underline{}$
   Subtract 1: $\underline{} - 1 = \underline{}$
   So, $6 + 3 = \underline{}$.

3. $5 + 7 = ?$
   I can use a doubles fact.
   $5 + \underline{} = \underline{}$
   Add 2 more: $\underline{} + 2 = \underline{}$
   So, $5 + 7 = \underline{}$.

Fluency Check

What is the sum or difference?

4. 6 + 7 = _____

5. 11 − 6 = _____

6. 8 + 9 = _____

7. 4 + 7 = _____

8. 3 + 6 = _____

9. 15 − 7 = _____

10. 5 + 3 = _____

11. 13 − 6 = _____

12. 14 − 8 = _____

13. 7 + 5 = _____

14. 5 + 6 = _____

15. 12 − 5 = _____

Fluency Talk

How can you make a 10 to help you add 7 + 5? Explain.

How can you use a doubles fact to subtract 17 − 8? Explain your thinking.

Strategies to Add 3-Digit Numbers

Focus Question

What strategies can I use to add 3-digit numbers?

Hi, I'm Riley.

I want to be an automotive engineer. A minivan goes 284 miles on one tank of gas. A small car goes 367 miles. I can use strategies to add to find the total number of miles!

Name _____

Greatest and Least Sums

Challenge 1

Find the greatest possible sum. Use one digit, from 1 through 9, in each box. Use each digit only once.

☐☐ + ☐☐ = ____

Challenge 2

Find the least possible sum. Use one digit, from 1 through 9, in each box. Use each digit only once.

☐☐ + ☐☐ = ____

Challenge 3

Find the greatest possible sum. Use one digit, from 1 through 9, in each box. Use each digit only once.

☐ + ☐☐ = ____

Lesson 9-1

Use Mental Math to Add 10 or 100

Be Curious

**What do you notice?
What do you wonder?**

Math is... Mindset

What helps you feel relaxed when you are frustrated?

Unit 9 • Strategies to Add 3-Digit Numbers

Learn

How can you help Clara complete the table?

141 + 10 = ?	141 + 100 = ?
161 + 10 = ?	161 + 100 = ?
197 + 10 = ?	197 + 100 = ?
297 + 10 = ?	297 + 100 = ?

When you add 10 to a number, the tens digit goes up by 1.

When you add 100 to a number, the hundreds digit goes up by 1.

Math is... Patterns

What do you notice about the digits in the ones place when you add 10 to a number?

If there are 9 tens, the tens digit changes to 0 and the hundreds digit goes up by 1.

141 + 10 = **151**	141 + 100 = **241**
161 + 10 = **171**	161 + 100 = **261**
197 + 10 = **207**	197 + 100 = **297**
297 + 10 = **307**	297 + 100 = **397**

You can use patterns to add 10 or 100 to 3-digit numbers.

Work Together

What is the sum?

385 + 10 = _____   385 + 100 = _____

493 + 10 = _____   493 + 100 = _____

690 + 10 = _____   690 + 100 = _____

88   Lesson 1 • Use Mental Math to Add 10 or 100

On My Own

Name _____

Is the statement true or false? Explain your answer.

1. The tens digit always goes up by 1 when you add 10 to a 3-digit number.

2. Addition patterns can help you add 10 or 100 to a 3-digit number.

What is the sum? Use a number line to show your work.

3. 382 + 10 = _____

4. 497 + 10 = _____

What is the sum?

5. 703 + 10 = _____

6. 894 + 10 = _____

7. 483 + 100 = _____

8. 350 + 100 = _____

9. **STEM Connection** Sienna keeps track of her steps. She takes 276 steps before breakfast. She takes 100 steps after breakfast. How many steps has she taken so far?

10. **Extend Your Thinking** Mikala has 757 pennies. Her brother gives her 10 pennies. Her sister gives her 100 pennies. How many pennies does Mikala have now?

How can you use patterns to mentally add 10 or 100?

Math is... Mindset

What has helped you feel relaxed when you are frustrated?

Lesson 9-2
Represent Addition with 3-Digit Numbers

Be Curious

**What do you notice?
What do you wonder?**

Day	Tacos Sold
Friday	73
Saturday	125
Sunday	112

Math is... Mindset

What helps you know when there is a problem?

Unit 9 • Strategies to Add 3-Digit Numbers

Learn

A taco shop owner records how many tacos she sells each day.

How many tacos did she sell on Saturday and Sunday?

Day	Tacos Sold
Friday	73
Saturday	125
Sunday	112

You can use base-ten blocks to represent 3-digit addition problems.

125 + 112 = ?

```
  125
+ 112
-----
  237
```

2 hundreds   3 tens   7 ones

The taco shop owner sold 237 tacos on Saturday and Sunday.

Math is... Choosing Tools

What other tool can you use to represent the problem?

One way to add 3-digit numbers is to add the ones, the tens, and then the hundreds.

Work Together

What is the sum?

243 + 146 = _____

92   Lesson 2 • Represent Addition with 3-Digit Numbers

On My Own

Name _____

What is the sum? Use base-ten shorthand to show your work.

1. 84 + 115 = ____

2. 206 + 481 = ____

Is the statement true or false? Circle the correct answer.

3. The number of hundreds in the sum of 243 + 125 is 6.

   True   False

4. The number of tens in the sum of 314 + 583 is 9.

   True   False

5. Win has 213 stickers. He buys 150 more stickers. How many stickers does Win have now?

6. Monique has 156 cards. She gets 42 more cards. How many cards does she have now?

7. **Error Analysis** Val writes 316 + 153 = 369. How do you respond to Val?

8. **Extend Your Thinking** How can you use base-ten blocks or base-ten shorthand to add 102 + 21 + 74?

 **Reflect**

How can representing each addend help you add 3-digit numbers?

Math is... Mindset

What has helped you know when there is a problem?

Lesson 9-3
Represent Addition with 3-Digit Numbers with Regrouping

Be Curious

Is this always true?

When adding tens, the number of hundreds and ones never change.

Math is... Mindset

What helps you understand thinking that is different from yours?

Learn

Dorian kept track of the number of people that visited the museum for 3 months.

Month	Visitors
April	242
May	292
June	228

How many people visited in May and June?

You can use base-ten blocks to represent 3-digit addition.

$$292 + 228 = ?$$

4 hundreds 11 tens 10 ones = 5 hundreds 2 tens

```
  292
+ 228
```

Math is... Exploring

How is regrouping tens similar to regrouping ones?

5 hundreds 2 tens = ?

520 people visited the museum in May and June.

When you add 3-digit numbers, sometimes you regroup 10 ones as 1 ten. Sometimes you regroup 10 tens as 1 hundred.

Work Together

How many people visited the museum in April and May?

On My Own

Name _____

1. Which equations need regrouping? Choose all the correct answers.

   A. 231 + 159 = ?

   B. 178 + 194 = ?

   C. 214 + 235 = ?

   D. 328 + 271 = ?

What is the sum? Show your work.

2. 195 + 265 = _____

3. 393 + 225 = _____

Unit 9 · Strategies to Add 3-Digit Numbers

4. The table shows the number of tickets sold on Saturday and Sunday. How many tickets are sold on both days?

Day	Tickets Sold
Saturday	219
Sunday	346

5. **STEM Connection** On Monday, C.J. surveyed 258 students about the new cafeteria menu items. On Tuesday, he surveyed 194 more students. How many students did C.J. survey in all?

6. **Extend Your Thinking** How can you regroup to find the sum? Explain.

$$173 + 126 + 249 = \_\_\_\_$$

 **Reflect**

How do you know when to regroup?

Math is... Mindset

What helped you understand thinking that is different from yours?

Lesson 9-4
Decompose Addends to Add 3-Digit Numbers

Be Curious

Which doesn't belong?

237 + 141

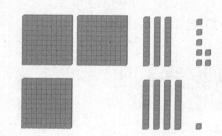

200 + 100 + 30 + 40 + 7 + 1

Math is... Mindset

What do you do well in math? In reading?

Learn

Pete is adding 237 + 189.

How might he decompose each addend?

You can decompose addends by place value to add.

237 + 189 = ?

237 → 200 + 30 + 7   +   189 → 100 + 80 + 9   = ?

Add by place value.

200 + 100 = 300
30 + 80 = 110
7 + 9 = 16

Add partial sums.

300 + 110 + 16 = **426**

237 + 189 = **426**

One strategy for adding 3-digit numbers is to decompose both addends by place value to find partial sums.

Math is... Structure

How can using partial sums help you add 3-digit numbers?

Work Together

How can you decompose both addends to add?

256 + 368 = _____

Lesson 4 • Decompose Addends to Add 3-Digit Numbers

On My Own

Name _____

1. Which expressions show both addends decomposed by place value? Choose all the correct answers.

   A.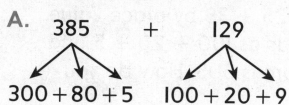

   B.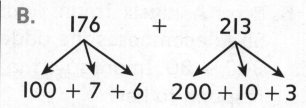

   C.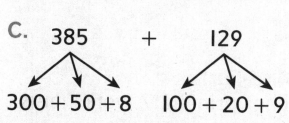

   D.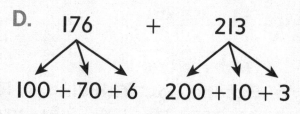

2. How can you decompose the addend by place value?

   a. 168 + 320 = ?

   ___ + ___ + ___   ___ + ___ + ___

   b. Add hundreds: ___ + ___ = ___

   Add tens: ___ + ___ = ___

   Add ones: ___ + ___ = ___

   c. Solve using partial sums:

   ___ + ___ + ___ = ___

Unit 9 • Strategies to Add 3-Digit Numbers

What is the sum? Decompose both addends to solve.

3. 143 + 286 = _____

4. 219 + 453 = _____

5. **Error Analysis** Imani adds 125 + 38 by place value. She decomposes the addends as 100 + 20 + 5 and 300 + 80. Imani says the sum is 505. How do you respond to her?

6. **Extend Your Thinking** A teacher prints 217 worksheets. Another teacher prints 196 worksheets. How many worksheets were printed? Explain your thinking.

Reflect

How does decomposing by place value help you add 3-digit numbers?

Math is... Mindset

How did your strengths in reading help you in math today?

Lesson 9-5

Decompose One Addend to Add 3-Digit Numbers

Be Curious

**How are they the same?
How are they different?**

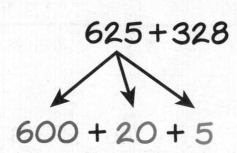

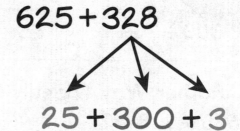

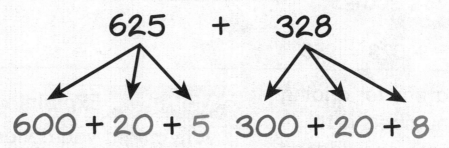

Math is... Mindset

What do you need to be ready to learn?

Unit 9 • Strategies to Add 3-Digit Numbers 103

Learn

Ms. Li's class is adding 625 + 328. Fran and Noel use different strategies to find the sum.

Can you use both strategies to find the sum of 625 + 328?

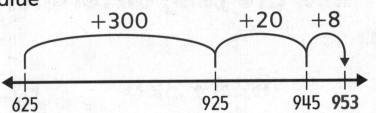

Fran's Strategy
625 + 300 + 20 + 8 = ?

Noel's Strategy
625 + 25 + 300 + 3 = ?

You can decompose addends in different ways.

▶ **One Way** Place Value

625 + 328 = ?
        ↙ ↓ ↘
    300 + 20 + 8

+300, +20, +8 on number line: 625 → 925 → 945 → 953

▶ **Another Way** Friendly Numbers

625 + 328 = ?
        ↙ ↓ ↘
    25 + 300 + 3

+25, +300, +3 on number line: 625 → 650 → 950 → 953

625 + 328 = **953**

One strategy for adding 3-digit numbers is to decompose one addend.

Math is... Explaining
Why might you decompose an addend in a different way?

Work Together

What is the sum? Decompose one addend to solve.

437 + 264 = _____

104 Lesson 5 • Decompose One Addend to Add 3-Digit Numbers

On My Own

Name _____

How can you decompose one addend? Choose all the correct answers.

1. 517 + 243 = ?
   A. 500 + 1 + 7 + 243
   B. 500 + 10 + 7 + 243
   C. 517 + 200 + 40 + 3
   D. 517 + 200 + 30 + 4

2. 495 + 378 = ?
   A. 495 + 300 + 7 + 8
   B. 495 + 300 + 70 + 8
   C. 400 + 50 + 9 + 378
   D. 400 + 90 + 5 + 378

What is the sum? Decompose one addend by place value.

3. a. 472 + 138 = ?

   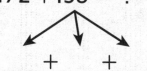

   ____ + ____ + ____

   b. Add: ____ + ____ + 30 + 8 = ____

   c. Solve: 472 + 138 = ____

4. a. 307 + 216 = ?

   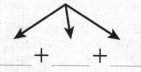

   ____ + ____ + ____

   b. Add: ____ + ____ + 10 + 6 = ____

   c. Solve: 307 + 216 = ____

What is the sum? Decompose one addend.
Use a number line to show your work.

5. 193 + 279 = _____

6. 340 + 156 = _____

7. **Extend Your Thinking** Xavier scored 273 points last season and 358 points this season. How many points did Xavier score in both seasons combined? Show your thinking.

Reflect

Why might you decompose one addend instead of both addends when adding?

Math is... Mindset

What helped you be ready to learn?

Lesson 9-6
Adjust Addends to Add 3-Digit Numbers

Be Curious

What do you notice?
What do you wonder?

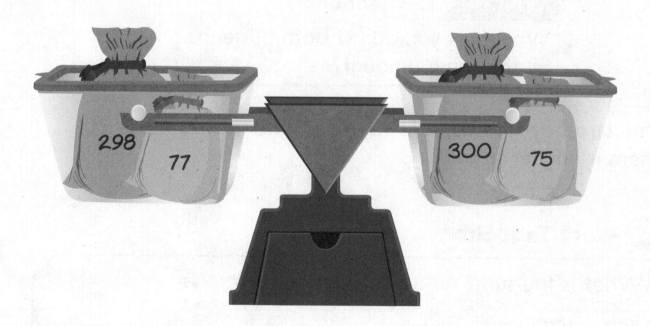

Math is... Mindset

What helps you understand how others are feeling?

Learn

Aubree needs to find the sum of 298 + 77.

How can she adjust the addends to make them easier to add?

▶ **One Way** Make 298 a friendly number.

298 + 77 = ?
+2   −2
↓   ↓
300 + 75 = 375

▶ **Another Way** Make 77 a friendly number.

298 + 77 = ?
−3   +3
↓   ↓
295 + 80 = 375

Math is... Connections

Why must you adjust both addends by the same amount?

One strategy for adding is to adjust addends to make them friendlier to add.

Work Together

What is the sum? Adjust the addends to solve.

349 + 168 = _____

108 Lesson 6 • Adjust Addends to Add 3-Digit Numbers

On My Own

Name _____

1. How can you adjust the addends? Choose all the correct answers.

$$554 + 397 = ?$$

A. $557 + 400$ B. $550 + 393$

C. $551 + 400$ D. $550 + 401$

How can you adjust addends to find the sum? Fill in the numbers.

2. $387 + 199 = $ ___
   ☐   ☐
   ↓   ↓
   ___ + ___ = ___

3. $267 + 525 = $ ___
   ☐   ☐
   ↓   ↓
   ___ + ___ = ___

4. $486 + 305 = $ ___
   ☐   ☐
   ↓   ↓
   ___ + ___ = ___

5. $175 + 203 = $ ___
   ☐   ☐
   ↓   ↓
   ___ + ___ = ___

6. What is the sum? Adjust the addends.

   597 + 290 = _____

7. **Error Analysis** Henley adjusted addends to find the sum of 227 + 198. How do you respond to her?

   Henley's work:
   227 + 198 = 429
   [+2]   [+2]
   ↓   ↓
   229 + 200 = 429

8. **Extend Your Thinking** Alyssa read for 158 minutes last week and 193 minutes this week. How many minutes did Alyssa read in all? Explain two ways to adjust the addends to solve.

Reflect

Why is the sum of adjusted addends the same as the sum of the original addends?

Math is... Mindset

What helped you understand how others are feeling?

Lesson 9-7
Explain Addition Strategies

Be Curious

What do you notice?
What do you wonder?

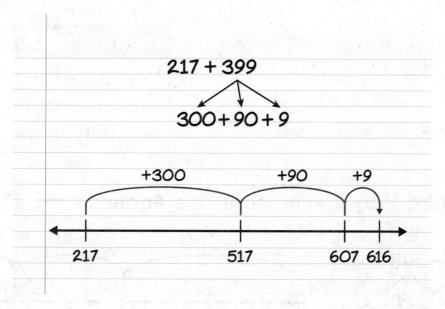

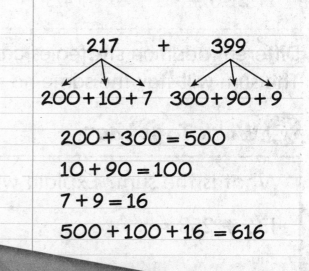

Math is... Mindset

Why is it important to speak clearly and concisely?

Unit 9 • Strategies to Add 3-Digit Numbers

Learn

What strategy would you use to find the total number of people at the soccer game?

423 people

398 people

▶ **One Way** Decompose Both Addends

423 + 398

400 + 20 + 3   300 + 90 + 8

400 + 300 = 700
20 + 90 = 110
3 + 8 = 11
700 + 110 + 11 = 821

▶ **Another Way** Adjust Addends

423 + 398 = ?
−2   +2
421 + 400 = 821

Math is... Explaining

Which strategy do you think is best for the numbers in the problem?

▶ **A Third Way** Decompose One Addend

423 + 398

300 + 90 + 8

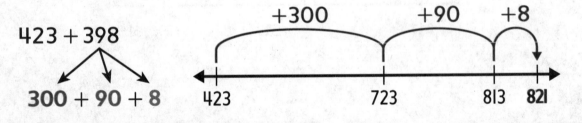

Different addition strategies can be used to add 3-digit numbers. The sum will stay the same no matter what strategy is used.

Work Together

What is the sum? Explain what strategy you used.

436 + 253 = _____

112   Lesson 7 • Explain Addition Strategies

On My Own

Name _____

What addition strategy is shown? Circle the correct answer.

1. $346 + 299 = 645$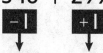
   $345 + 300 = 645$

   A. adjust addends
   B. decompose both addends
   C. decompose one addend
   D. skip counting

2.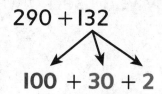

   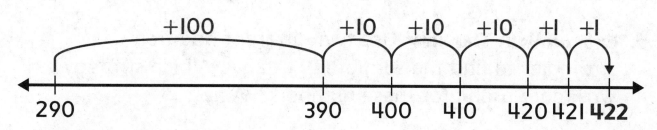

   A. adjust addends
   B. decompose both addends
   C. decompose one addend
   D. make a 10

3.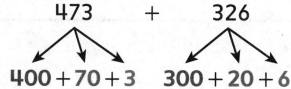

   $400 + 300 = 700$
   $70 + 20 = 90$
   $3 + 6 = 9$
   $700 + 90 + 9 = 799$

   A. adjust addends
   B. decompose both addends
   C. decompose one addend
   D. skip counting

Unit 9 • Strategies to Add 3-Digit Numbers

4. Marcus has 437 dimes and Florentine has 246 dimes. How many dimes do they have in all? Explain your thinking.

5. **STEM Connection** Deven mixed 427 minutes of music and 508 minutes of nature sounds. How many minutes of audio did he mix? Explain what strategy you used and why.

6. **Extend Your Thinking** Use two different addition strategies to find the sum of 129 + 287. Which strategy was more useful for these numbers? Why?

Reflect

Why is it helpful to know different addition strategies?

Math is... Mindset
What helped you to speak clearly and concisely?

Unit 9
Addition Problems

Name _____

**Solve the problem.
Decide if the answer is more than the bold number shown.**

1. Lily has 258 nickels and 129 dimes. How many coins does she have?

   Is the answer **more than 380?**

   Circle Yes or No.

   Yes   No

   Explain why you chose Yes or No.

---

2. 397 tickets were sold this week. 113 tickets were sold last week. How many tickets were sold?

   Is the answer **more than 500?**

   Circle Yes or No.

   Yes   No

   Explain why you chose Yes or No.

Solve the problem.
Decide if the answer is more than the bold number shown.

3. A factory made 436 shirts and some jackets. They made 126 fewer shirts than jackets. How many jackets did they make?

   Is the answer **more than 565?**

   Circle Yes or No.

   Yes		No

   Explain why you chose Yes or No.

Reflect On Your Learning

Unit Review   Name _____

Vocabulary Review

Use the vocabulary to complete each sentence.

> adjust   decompose
> hundreds   partial sums
> friendly numbers

1. When adding _____ you add the digits in one place value at a time, and then add those sums to find the total sum. (Lesson 9-4)

2. You _____ to make an equation easier to solve. (Lesson 9-6)

3. You _____ a number by breaking it into different parts. (Lesson 9-5)

4. Numbers that are easy to add are _____. (Lesson 9-3)

5. In the number 234, 2 is in the _____ place. (Lesson 9-1)

Unit 9 • Strategies to Add 3-Digit Numbers   117

Review

6. What is the sum of 592 + 135? Use place value to decompose each addend. Then add the partial sums.

   (Lesson 9-4)

	Hundreds	Tens	Ones
592			
135			

   hundreds: _____ + _____ = _____

   tens: _____ + _____ = _____

   ones: _____ + _____ = _____

   partial sums: _____ + _____ + _____ = _____

7. Mariah earns 256 points. Cody earns 398 points. How can you adjust the addends to make it easier to find the total number of points they earned? Choose all the correct answers. (Lesson 9-6)

   A. Add 2 to 398. Add 2 to 256.

   B. Add 2 to 398. Subtract 2 from 256.

   C. Add 4 to 256. Add 4 to 398.

   D. Add 4 to 256. Subtract 4 from 398.

8. What is the sum? Use patterns to help you add.

   (Lesson 9-1)

   504 + 10 = _____

9. What is the sum? Use base-ten shorthand to show your work. (Lesson 9-2)

633 + 145 = _____

10. What is the sum? Use patterns to help you add. (Lesson 9-1)

278 + 100 = _____

11. What is the sum? Use base-ten shorthand to show your work. (Lesson 9-3)

454 + 377 = _____

12. What is the sum? Decompose the second addend to find the sum. (Lesson 9-5)

547 + 158 = ?

158 = _____ + 50 + 8

547 + _____ + 50 + 8 = _____

Unit 9 • Strategies to Add 3-Digit Numbers   119

Performance Task

An automotive engineer recorded the number of cars made at 4 companies.

Company A	Company B	Company C	Company D
231	325	194	337

Part A: How many cars were made at Company A and Company B?

Part B: How many cars were made at Company B and Company C?

Part C: How many cars were made at Company C and Company D?

What strategies can you use to add 3-digit numbers?

Unit 9
Fluency Practice

Name _____

Fluency Strategy

You can use base-ten blocks to help add tens to a number.
14 + 20 = ?

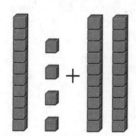

There are 3 tens and 4 ones.
So, 14 + 20 = **34**.

You can use base-ten blocks to help subtract tens from a number.
67 − 40 = ?

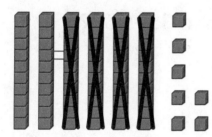

There are 2 tens and 7 ones left.
So, 67 − 40 = **27**.

Fluency Flash

What is the sum or difference?

1. 51 − 30 = _____

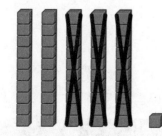

2. 39 + 40 = _____

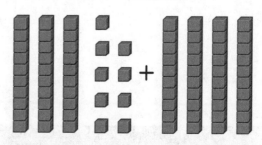

Fluency Check

What is the sum or difference?

3. 23 + 50 = _____

4. 14 − 8 = _____

5. 77 − 40 = _____

6. 34 − 10 = _____

7. 4 + 7 = _____

8. 6 + 3 = _____

9. 46 + 10 = _____

10. 3 + 5 = _____

11. 11 − 5 = _____

12. 24 + 60 = _____

13. 55 − 30 = _____

14. 17 − 9 = _____

Fluency Talk

How can you use base-ten blocks to add 35 + 50? Explain.

How can you use a known fact to add 7 + 5? Explain.

Unit 10

Strategies to Subtract 3-Digit Numbers

Focus Question

What strategies can I use to subtract 3-digit numbers?

Hi, I'm Kayla.

I want to be a landscape architect. I plan to design two parks across the street from each other. I want to find out how many more people can be at one park than the other. I can subtract to find this out.

STEM video | GO ONLINE

Name _____

Greatest and Least Differences

Challenge 1

Find the greatest possible difference. Use one digit, from 1 through 9, in each box. Use each digit only once.

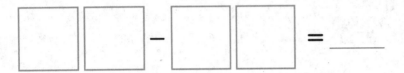

Challenge 2

Find the least possible difference. Use one digit, from 1 through 9, in each box. Use each digit only once.

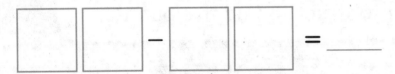

Challenge 3

Make a difference of 5. Use one digit, from 1 through 9, in each box. Use each digit only once.

☐☐ − ☐ = 5

Lesson 10-1

Use Mental Math to Subtract 10 or 100

Be Curious

What do you notice?
What do you wonder?

Math is... Mindset

What do you do to control your actions in class?

Unit 10 • Strategies to Subtract 3-Digit Numbers

Learn

Robert is working to complete the table.

What patterns do you notice that can help you subtract 10 or 100?

538 − 10 = 528	538 − 100 = 438
528 − 10 = ?	438 − 100 = ?
518 − 10 = ?	338 − 100 = 238
508 − 10 = 498	238 − 100 = ?
498 − 10 = ?	138 − 100 = ?

Subtracting 10 makes the tens digit go down by 1.

Subtracting 100 makes the hundreds digit go down by 1.

If there are 0 tens, the tens digit changes to 9 and the hundreds digit goes down by 1.

538 − 10 = 528	538 − 100 = 438
528 − 10 = 518	438 − 100 = 338
518 − 10 = 508	338 − 100 = 238
508 − 10 = 498	238 − 100 = 138
498 − 10 = 488	138 − 100 = 38

Math is... Patterns

How are the patterns similar? How are they different?

You can use patterns to subtract 10 or 100 from 3-digit numbers.

Work Together

What is the difference?

754 − 10 = _____

551 − 10 = _____

303 − 10 = _____

925 − 100 = _____

407 − 100 = _____

185 − 100 = _____

On My Own

Name _____

1. Which equations are true? Choose all the correct answers.

   A. 600 − 10 = 50
   B. 600 − 10 = 590
   C. 500 − 100 = 600
   D. 600 − 100 = 500

What is the difference? Use the number line to show your work.

2. 908 − 10 = _____

3. 189 − 100 = _____

What is the difference?

4. 285 − 10 = _____
5. 717 − 10 = _____
6. 804 − 100 = _____
7. 198 − 100 = _____

8. Amari does 85 push-ups. Evan does 10 fewer push-ups than Amari. How many push-ups does Evan do?

9. **STEM Connection** Kayla is working with a team to plant 772 trees at a park. They already planted 100 trees. How many trees do Kayla and the team have left to plant?

10. **Extend Your Thinking** Stephanie has 127 dollar bills. She puts 100 dollar bills in the bank. Then she gives her sister 10 dollar bills. How many dollar bills does Stephanie have now?

Reflect

How can you use patterns to mentally subtract 10 or 100 from a 3-digit number?

Math is... Mindset

What helped you control your actions in class?

Lesson 10-2
Represent Subtraction with 3-Digit Numbers

Be Curious

What do you notice?
What do you wonder?

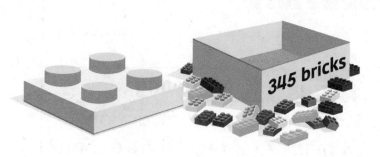

Math is... Mindset
How well do you think you will do with today's tasks?

Learn

Carmen has a box of 345 building bricks. She uses 114 bricks to build a house.

How many building bricks are left?

Show 345. Then subtract 114.

345 − 114 = ?

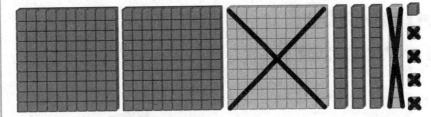

345 − 114 = **231**

There are 231 bricks left.

You can use base-ten blocks to represent and solve 3-digit subtraction equations.

Math is... Modeling

How is 3-digit subtraction similar to 2-digit subtraction?

 Work Together

Myles has 275 stamps in his stamp collection. He gives 132 stamps to his sister. How many stamps are left in Myles's collection? Use base-ten shorthand to show your work.

Lesson 2 • Represent Subtraction with 3-Digit Numbers

On My Own

Name _____

Which equation is represented by the base-ten blocks? Choose the correct answer.

1.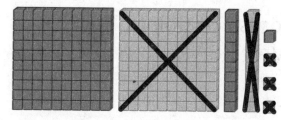

   A. 221 + 113 = 334
   B. 224 − 113 = 111
   C. 224 + 113 = 337
   D. 221 − 113 = 108

2.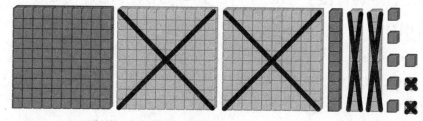

   A. 338 − 202 = 136
   B. 335 − 222 = 113
   C. 338 − 222 = 116
   D. 328 − 202 = 126

What is the difference? Use base-ten shorthand to show your work.

3. 279 − 157 = _____

4. 386 − 105 = _____

Unit 10 • Strategies to Subtract 3-Digit Numbers

Represent the problem using base-ten shorthand.

5. Mateo has 725 football cards. He gives away 205 cards. How many football cards does Mateo still have?

6. Molly scores 365 points. She loses 124 points. How many points does Molly have left?

7. **Error Analysis** Hank writes 416 − 105 = 301. How do you respond to him?

8. **Extend Your Thinking** How can you use base-ten blocks to solve 256 − 134?

Reflect

How can base-ten blocks help you subtract 3-digit numbers?

Math is... Mindset

How well do you think you did with today's tasks?

Lesson 10-3
Decompose One 3-Digit Number to Count Back

Be Curious

How are they the same? How are they different?

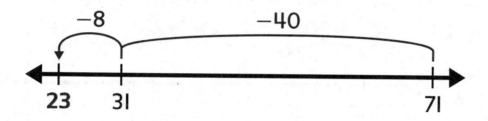

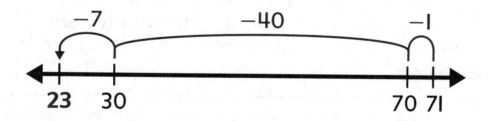

Math is... Mindset

What are some ways to resolve disagreements with your classmates?

Unit 10 • Strategies to Subtract 3-Digit Numbers 133

Learn

Mary and Juan will decompose 257 to subtract.

$431 - 257 = ?$

How can they decompose 257 to subtract?

▶ **One Way** Decompose by place value.

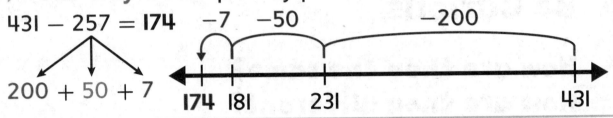

$431 - 257 = 174$
$200 + 50 + 7$

▶ **Another Way** Decompose to make friendly numbers.

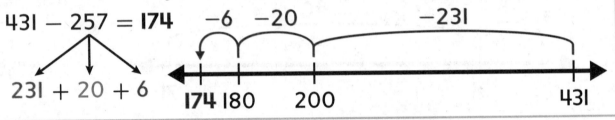

$431 - 257 = 174$
$231 + 20 + 6$

One strategy for subtracting 3-digit numbers is to decompose one number and count back on a number line.

> **Math is... Thinking**
>
> Which way of decomposing was more efficient for you?

Work Together

How can you decompose to find the difference? Show the subtraction on the number line.

$652 - 234 = $ _____

____ + ____ + ____

134 Lesson 3 • Decompose One 3-Digit Number to Count Back

On My Own

Name _____

How can you decompose the bold number?
Circle the correct answer.

1. 319 − **127** = ?

    100 + 20 + 7   120 + 70

2. 405 − **169** = ?

    16 + 90   105 + 60 + 4

3. 428 − **290** = ?

    200 + 9   200 + 90

4. 516 − **320** = ?

    300 + 2   316 + 4

How can you decompose to find the difference?
Show the subtraction on the number line.

5. 413 − 256 = _____

    256 = ____ + ____ + ____

6. 614 − 388 = _____

    388 = ____ + ____ + ____

Unit 10 · Strategies to Subtract 3-Digit Numbers

7. **STEM Connection** Riley's dad can drive his car 383 miles on a full tank of gas. Riley's mom can drive her car 500 miles on a full tank of gas. How many more miles can her mom drive her car than her dad on a full tank of gas?

8. **Extend Your Thinking** Decompose by place value and another way to find the difference of 469 − 275. Which way is more efficient for you? Explain.

 **Reflect**

How can you decompose a 3-digit number to help you subtract?

Math is... Mindset

What helped you resolve disagreements with your classmates?

Lesson 10-4
Count On to Subtract 3-Digit Numbers

Be Curious

Tell me everything you can.

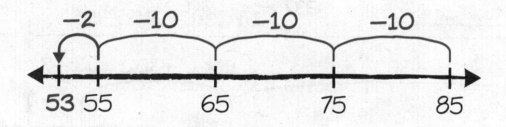

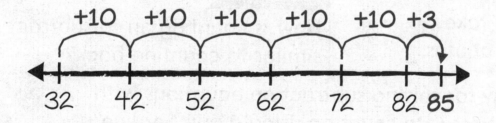

Math is... Mindset

What behaviors show that you respect your classmates?

Learn

Tasha has space for 965 photos on her camera. She has already taken 628 photos.

How many more photos can Tasha take before her camera runs out of space?

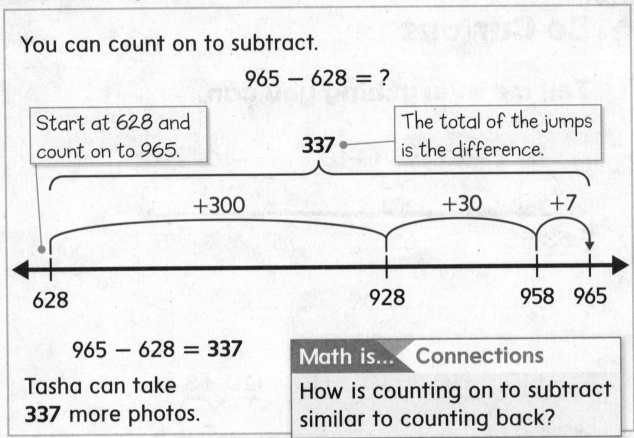

You can count on to subtract.

$$965 - 628 = ?$$

Start at 628 and count on to 965.

337

The total of the jumps is the difference.

$965 - 628 = 337$

Tasha can take **337** more photos.

Math is... Connections

How is counting on to subtract similar to counting back?

One strategy for solving subtraction equations with 3-digit numbers is to count on using a number line.

Work Together

What is the difference? Use the number line to count on.

$518 - 343 =$ _____

138 Lesson 4 • Count On to Subtract 3-Digit Numbers

On My Own

Name _____

1. Which equation is related to 419 − 158? Choose the correct answer.

   A. 158 + 419 = ?

   B. 158 + ? = 419

   C. ? − 158 = 419

   D. 419 + 158 = ?

How can you count on to subtract? Complete the number line and find the difference.

2. 333 − 212 = ____

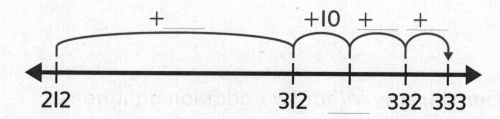

3. 354 − 228 = ____

4. 671 − 352 = ____

Unit 10 • Strategies to Subtract 3-Digit Numbers

5. Marissa has 356 e-mail messages in her inbox. She deletes 108 e-mail messages. How many e-mail messages are left in Marissa's inbox? Write an equation to represent the problem. Use the number line to count on.

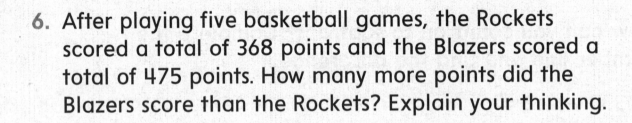

6. After playing five basketball games, the Rockets scored a total of 368 points and the Blazers scored a total of 475 points. How many more points did the Blazers score than the Rockets? Explain your thinking.

7. **Extend Your Thinking** What two addition equations are related to 283 − 157? Explain how you can use addition to find the difference.

 **Reflect**

How can you count on to subtract 3-digit numbers?

Math is... Mindset

How has your behavior shown that you respect your classmates?

Lesson 10-5
Regroup Tens

Be Curious

How are they the same? How are they different?

265 − 41

265 − 48

265 − 131

265 − 139

Math is... Mindset

What helps you stay focused on your work?

Unit 10 • Strategies to Subtract 3-Digit Numbers

Learn

The mail carrier had 546 letters. She delivered 128 letters.

How many letters does the mail carrier still need to deliver?

Show 546 with base-ten blocks.

546 − 128 = ?

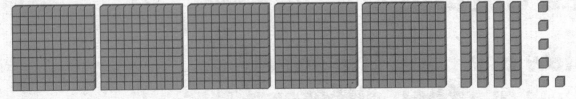

Decompose a ten to subtract.

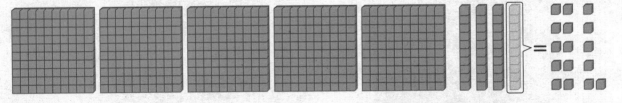

Then subtract 128.

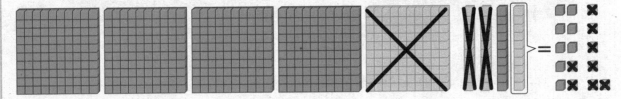

546 − 128 = **418**

The mail carrier has 418 letters to deliver.

You may need to regroup a ten when subtracting 3-digit numbers.

Math is... Thinking

Why is the order in which you subtract important?

 Work Together

Rami has 851 files on his computer. He deletes 545 files. How many files are left on his computer?

142　Lesson 5 • Regroup Tens

On My Own

Name _____

Is regrouping needed to subtract? Circle the correct answer.

1. 172 − 45

   Yes   No

2. 456 − 234

   Yes   No

3. 728 − 204

   Yes   No

4. 598 − 379

   Yes   No

What is the difference? Use base-ten shorthand to show your work.

5. 364 − 138 = _____

6. 234 − 125 = _____

7. Davis has a box of 842 photos. He puts 426 photos in albums. How many photos are still in the box?

    842 − 426 = ?   40 − 20 = 20
                                                    424
    800 − 400 = 400   2 − 6 = 4

8. **Error Analysis** Roger writes this equation 573 − 245 = 338. How do you respond to Roger?

9. **Extend Your Thinking** Vera has 264 beads. She uses 147 beads. Explain why regrouping is needed to find how many beads Vera has left.

Reflect

How do you know when you need to regroup?

Math is... Mindset

What has helped you stay focused on your work?

Lesson 10-6

Regroup Tens and Hundreds

Be Curious

**What do you notice?
What do you wonder?**

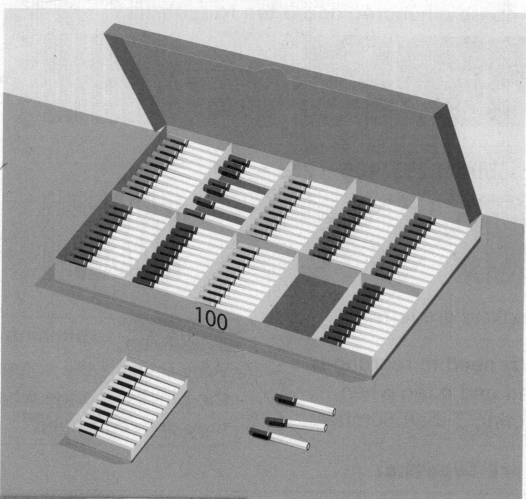

100

Math is... Mindset

What helps you solve a problem?

Unit 10 • Strategies to Subtract 3-Digit Numbers   145

Learn

Some students take 253 of the markers.

How many markers are left?

Show 400 with base-ten blocks. 400 − 253 = ?

Decompose a hundred and a ten to subtract.

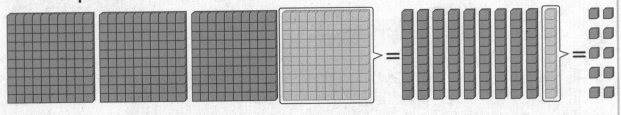

Then subtract 253. 400 − 253 = **147**

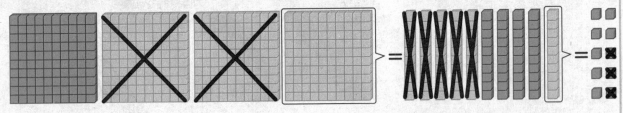

147 makers are left.

You may need to regroup a hundred and a ten when subtracting 3-digit numbers.

Math is... Explaining

Why doesn't the value of the blocks change when they are regrouped?

Work Together

There are 365 days in a year. Beck goes to school for 172 days. How many days does Beck not go to school?

On My Own

Name _____

How can you subtract 157 from the base-ten blocks? Circle Yes or No.

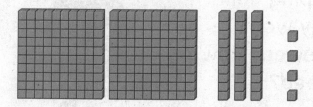

1. Do you need to regroup the hundreds?

   Yes   No

2. Do you need to regroup the tens?

   Yes   No

What is the difference? Use base-ten shorthand to show your work.

3. 428 − 149 = _____

4. 365 − 283 = _____

5. Trevor bakes 225 muffins for a bake sale. He sells 186 muffins. How many muffins does Trevor still have?

6. **STEM Connection** Kayla is helping her dad landscape their yard. They want 500 flowers. They have 367 flowers. How many more flowers do they need?

7. **Extend Your Thinking** Ian is driving to visit his family who lives 747 miles away. He stops for gas after 468 miles. How many more miles does Ian have left to drive? Explain why regrouping is needed to find the answer.

Reflect

How can you regroup tens and hundreds to subtract 3-digit numbers?

Math is... Mindset

What has helped you solve a problem?

Lesson 10-7
Adjust Numbers to Subtract 3-Digit Numbers

Be Curious

How are they the same?
How are they different?

$$498 - 251$$

$$497 - 250$$

$$500 - 253$$

Math is... Mindset
What helps you make good decisions?

Learn

How many pennies are left in Camila's piggy bank?

251 pennies

Camila takes out 197 pennies.

▶ **One Way** Make 197 a friendly number.

251 − 197 = ?

254 − 200 = **54**

Camila has 54 pennies left in her piggy bank.

▶ **Another Way** Make 251 a friendly number.

251 − 197 = ?
−1   −1
250 − 196 = **54**

Math is... Thinking

Why must you use the same operation to adjust both numbers?

One strategy for subtracting 3-digit numbers is to adjust numbers to make them friendlier to subtract.

Work Together

What is the difference? Adjust the numbers to solve.

349 − 173 = _____

On My Own

Name _____

1. How can you adjust the numbers to subtract? Choose all the correct answers.

   347 − 152 = ?

   A. 350 − 155

   B. 350 − 149

   C. 349 − 150

   D. 345 − 150

How can you adjust the numbers to find the difference? Fill in the numbers.

2. 259 − 47 = ?

   ☐ ☐
   ↓ ↓
   ___ − ___ = ___

3. 324 − 113 = ?

   ☐ ☐
   ↓ ↓
   ___ − ___ = ___

4. 415 − 298 = ?

   ☐ ☐
   ↓ ↓
   ___ − ___ = ___

5. 587 − 129 = ?

   ☐ ☐
   ↓ ↓
   ___ − ___ = ___

Unit 10 • Strategies to Subtract 3-Digit Numbers

6. Emilio goes to his grandmother's house that is 683 meters away. He sprints 328 meters and jogs the rest. How many meters does Emilio jog? Write an equation with friendly numbers to solve.

7. **Error Analysis** Deanna is finding the difference of 264 − 106 by adjusting the numbers to 260 − 110. How do you respond to Deanna?

8. **Extend Your Thinking** Mr. Park writes 298 − 143 = ?. Some students adjust the numbers to 300 − 145 and some adjust the numbers to 295 − 140. Which way of adjusting do you think is more efficient? Explain.

Reflect

Why is the difference of an adjusted equation the same as the difference of the original equation?

Math is... Mindset

What helped you make good decisions?

Explain Subtraction Strategies

Lesson 10-8

Be Curious

**What do you notice?
What do you wonder?**

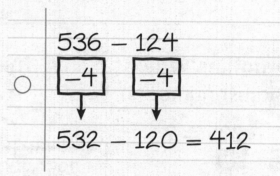

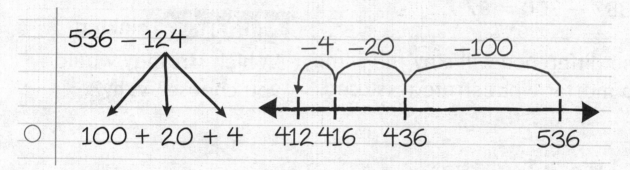

Math is... Mindset

What are some ways you can connect with your classmates?

Learn

How many more bottles of water were sold than cartons of milk?

Drink	Number Sold
water	382
milk	295

▶ **One Way** Decompose one number and count back.

382 − 295 = **87**

200 + 90 + 5

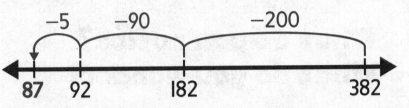

▶ **Another Way** Adjust numbers.

382 − 295 = ?
+5   +5
387 − 300 = **87**

▶ **A Third Way** Count on.

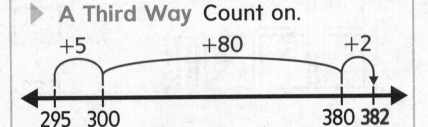

The difference will stay the same no matter what strategy is used.

Math is... Thinking
Which strategy would you choose? Why?

Work Together

What is the difference? Use a subtraction strategy. Then explain why you chose that strategy.

815 − 264 = _____

Lesson 8 • Explain Subtraction Strategies

On My Own

Name _____

Fill in the correct answer to complete the sentence.

1. To count on to find the difference of 493 − 217, start at _____.

2. To count back to find the difference of 872 − 549, start at _____.

Choose all the correct answers.

3. How can you adjust the numbers to find the difference?
   253 − 151 = ?
   A. 254 − 150
   B. 250 − 148
   C. 250 − 154
   D. 252 − 150

4. How can you decompose 325 to find the difference?
   523 − 325 = ?
   A. 32 + 5
   B. 300 + 2 + 5
   C. 300 + 20 + 5
   D. 300 + 20 + 3 + 2

5. Which equation is related to 928 − 499?
   A. 499 + ? = 928
   B. ? − 499 = 928
   C. 928 + 499 = ?
   D. 928 − ? = 499

Unit 10 • Strategies to Subtract 3-Digit Numbers   155

Use a subtraction strategy to solve. Then explain the subtraction strategy you used.

6. 867 − 189 = _____

7. Hallie has 500 blocks. 268 of the blocks are red. How many blocks are not red?

8. **Extend Your Thinking** Juan wants to sell 364 tickets to a school play. He already sold 198 tickets. How many tickets does Juan have left to sell? Use two different subtraction strategies to solve and explain which strategy is more efficient for you.

Reflect

Why is it helpful to know how to use different subtraction strategies?

Math is... Mindset

What helped you connect with your classmates?

Lesson 10-9
Solve Problems Involving Addition and Subtraction

Be Curious

What's the question?

There is a stack of maps at the zoo. Matt hands out some of the maps and Albert hands out some of the maps.

Math is... Mindset

What makes you feel excited in math?

Learn

There is a stack of 500 maps at the zoo. Matt hands out 284 maps and Albert hands out 115 maps.

How many maps are left?

Some problems have more than one question to answer.

How many maps do Matt and Albert hand out?

$284 + 115 = ?$

You can add to find the answer.

$284 + 115 = \mathbf{399}$

Matt and Albert hand out 399 maps.

How many maps are left?

$500 - 399 = ?$

You can subtract to find the answer.

$500 - 399 = 101$ — Think: $501 - 400 = ?$

There are 101 maps left.

You can use addition and subtraction to solve one- and two-step problems.

Math is... Planning

What strategies can you use to solve the problem?

Work Together

Zoe has 350 stamps. She uses 220 of the stamps. Then she buys 125 more stamps. How many stamps does Zoe have now?

On My Own

Name _____

1. Elaine has 294 buttons in a box. She gets 175 more buttons. How many buttons does Elaine have now?

   Which equation can you use to represent the word problem? Choose the correct answer.

   A. 294 − 175 = ?   B. 175 + ? = 294

   C. 294 + 175 = ?   D. 294 − ? = 175

Write an equation to represent the problem. Use any strategy to solve.

2. Jim has 461 bags of soil. He uses 286 bags. He buys 318 bags. How many bags of soil does Jim have now?

3. Stasia has 463 books. Troy has 159 fewer books. How many books does Troy have?

4. A scientist has 562 beakers and buys 185 new beakers. How many beakers does the scientist have altogether?

5. There are 247 blue pens in the drawer. There are 101 fewer red pens than blue pens. How many pens are in the drawer? Explain your thinking.

6. Mia scores 164 more points than Noah in the video game. Wyatt scores 123 fewer points than Mia. How many points does Wyatt score? Explain your thinking.

7. **Extend Your Thinking** Write a problem that has more than one question to answer using 3-digit numbers that involves addition and subtraction. Solve the problem using any strategy.

Reflect

What strategies can you use to solve problems with addition and subtraction?

Math is... Mindset

What has made you feel excited in math?

Unit 10
Addition and Subtraction Problems

Name _____

1. Mr. B's and Mrs. Yu's classes had a contest. Mr. B's class read 318 books. Mrs. Yu's class read 109 more books than that. How many books did Mrs. Yu's class read?

   Solve the problem.

   Circle the correct equation.

   a. 318 + 109 = ?
   b. 318 − 109 = ?
   c. 109 − 318 = ?

   Explain your choice.

2. A theater sold 327 tickets on Sunday. This was 119 fewer tickets than they sold on Saturday. How many tickets did they sell on Saturday?

   Solve the problem.

   Circle the correct equation.

   a. 327 − 119 = ?
   b. 119 − 327 = ?
   c. 327 + 119 = ?

   Explain your choice.

3. Sofia traveled 547 miles on Day 1. She traveled some more miles on Day 2. She traveled 687 miles in all. How many miles did she travel on Day 2?

Solve the problem.

Circle the correct equation.

a. $547 + 687 = ?$

b. $547 - 687 = ?$

c. $687 - 547 = ?$

Explain your choice.

Reflect On Your Learning

Unit Review

Name _____

Vocabulary Review

Use the vocabulary to complete each sentence.

adjust decompose
friendly numbers hundreds
regroup

1. Numbers that are easy to add or subtract are called _____. (Lesson 10-7)

2. In the number 892, 8 is in the _____ place. (Lesson 10-1)

3. You _____ numbers by adding the same amount to both numbers or subtracting the same amount from both numbers to make the numbers easier to subtract. (Lesson 10-7)

4. You _____ a number by breaking it into different parts by place value. (Lesson 10-3)

5. To take apart a ten or a hundred to show a number in a new way means you _____. (Lesson 10-5)

Review

6. What is the difference? (Lesson 10-6)

    563 − 295 = _____

7. How can you count on to subtract 524 − 383? Choose the correct answer. (Lesson 10-4)

    A.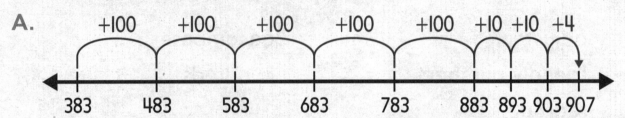

    B.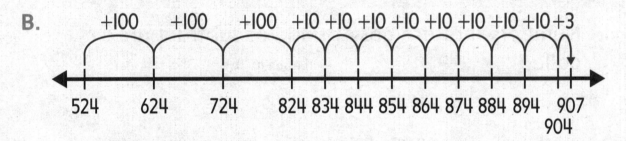

    C.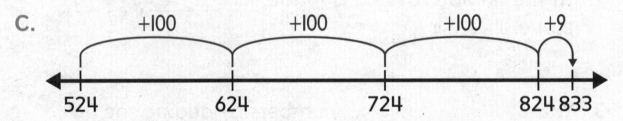

    D.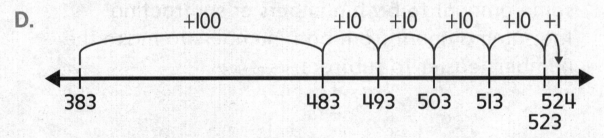

8. What is the difference? (Lesson 10-1)

    602 − 10 = _____

9. Jose asked 315 people to vote for their favorite color. There are 128 votes for red, 154 votes for green, and the rest of the votes are for blue. How many votes are for blue? (Lesson 10-9)

    _____ votes

10. What is the difference? (Lesson 10-2)

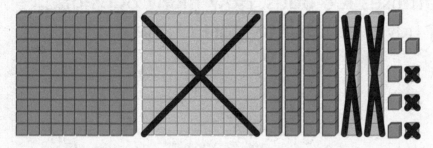

269 − 123 = _____

11. James counts back on the number line to find the difference of 822 − 142. Fill in the missing numbers to help James find the difference. What is the difference? (Lesson 10-3)

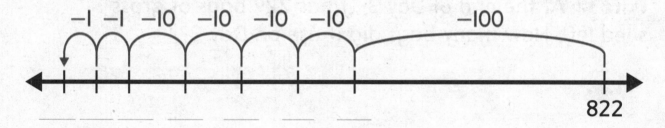

822 − 142 = _____

12. How can you adjust numbers to make friendly numbers to subtract 681 − 392? Choose the correct answer. (Lesson 10-7)

   A. Subtract 2 from 392. Subtract 2 from 681.

   B. Add 2 to 392. Subtract 2 from 681.

   C. Add 2 to 392. Add 3 to 681.

   D. Subtract 2 from 392. Add 2 to 681.

13. What is the difference? (Lesson 10-5)

   572 − 129 = _____

Unit 10 • Strategies to Subtract 3-Digit Numbers   165

Performance Task

A landscaping company buys 744 bags of grass seed.

Part A: On Day 1, it uses 106 bags. How many bags of grass seed does it have left for the start of the next day?

Part B: On Day 2, it uses 40 more bags than it did the day before. How many bags will it have left for the start of Day 3?

Part C: At the end of Day 3, it has 229 bags of grass seed left. How many bags did it use on Day 3?

 Reflect

What strategies can you use to subtract 3-digit numbers?

Unit 10
Fluency Practice

Name _____

Fluency Strategy

> You can decompose a number by making a ten to help you add or subtract.
>
> $37 + 5 = ?$ $\quad\quad\quad\quad\quad\quad$ $72 - 9 = ?$
> $\quad\quad\;\;$ 3 + 2 $\quad\quad\quad\quad\quad\quad\quad\quad\;\;$ 2 + 7
>
> $37 + 3 = 40$ $\quad\quad\quad\quad\quad\quad$ $72 - 2 = 70$
> $40 + 2 = 42$ $\quad\quad\quad\quad\quad\quad$ $70 - 7 = 63$
> So, $37 + 5 = 42$. $\quad\quad\quad\;\;$ So, $72 - 9 = 63$.

1. How can you decompose a number to make a ten to subtract $31 - 4$? Explain.

Fluency Flash

What is the sum or difference?

2. $83 - 8 =$ _____ $\quad\quad\quad\quad\quad$ 3. $48 + 7 =$ _____
 $\quad\;\;$ 3 + 5 $\quad\quad\quad\quad\quad\quad\quad\quad\;\;$ 2 + 5

Unit 10 • Strategies to Subtract 3-Digit Numbers

Fluency Check

What is the sum or difference?

4. 27 + 8 = _____

5. 22 − 10 = _____

6. 6 + 3 = _____

7. 67 + 9 = _____

8. 7 + 5 = _____

9. 56 + 7 = _____

10. 17 + 40 = _____

11. 4 + 7 = _____

12. 41 − 6 = _____

13. 78 − 20 = _____

14. 34 + 60 = _____

15. 71 + 10 = _____

Fluency Talk

How can you decompose a number to make a ten to add 89 + 7? Explain.

How can you use base-ten blocks to subtract 65 − 30? Explain.

Unit 11

Data Analysis

Focus Question

How can picture graphs, bar graphs, and line plots help me interpret data?

Hi, I'm Hugo.

I want to be a meteorologist. I can tell from a graph how many rainy days we have each month. Understanding graphs will be an important part of my job.

169

Name _____

Mystery Data

_____

	3	0	6	7
	4	5	1	5
	1	4	2	3
	5	2	6	3
	3	4	2	1

1. What do you notice about the information shown in the table?

_____

_____

_____

_____

2. What do you wonder about the information?

_____

_____

_____

_____

170 Ignite! • Mystery Data

Lesson 11-1
Understand Picture Graphs

Be Curious

**What do you notice?
What do you wonder?**

Math is... Mindset

What helps you make sense of a situation?

Unit 11 · Data Analysis 171

Learn

Izzy asks some students about their favorite subject. The **tally chart** shows the information she collected.

What subject is the most common?

Favorite Subject	
Subject	Tally
Math	||||‌ ||||‌
Reading	||||‌ ||||
Science	||||‌ |||

You can show the information, or the **data**, in a **picture graph**. Each picture shows one data point.

Favorite Subject

Math — 10 book pictures
Reading — 9 book pictures
Science — 8 book pictures

Each picture = 1 vote

- title
- There are 10 pictures in the Math row.
- key
- Each row has a label that names the **category**.

Math is the most common favorite subject.

Math is... Modeling

Why is it helpful to display the data in this way?

Drawing picture graphs can be a useful way to display data.

Work Together

How can you show the data using a picture graph?

Favorite Season	
Season	Tally
Spring	||||‌ ||
Fall	||||‌	
Summer	||||‌ |||
Winter	|||

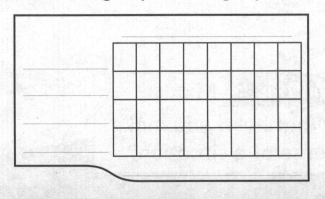

172 Lesson 1 • Understand Picture Graphs

On My Own

Name _____

How can you represent the data using a picture graph? Use the tally chart to make a picture graph.

1.

Favorite Sport	
Sport	Tally
Baseball	IIII
Football	IIII III
Basketball	IIII I
Soccer	II

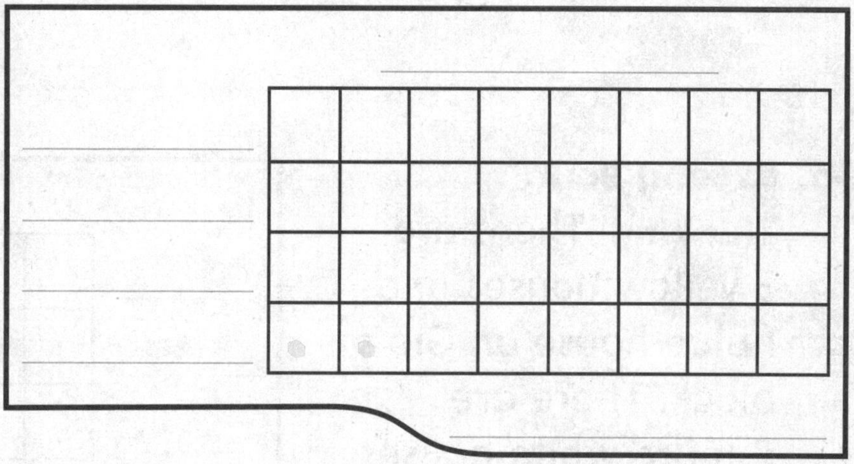

2.

Favorite Fruit	
Fruit	Tally
Banana	II
Apple	IIII I
Grapes	IIII I
Pear	III

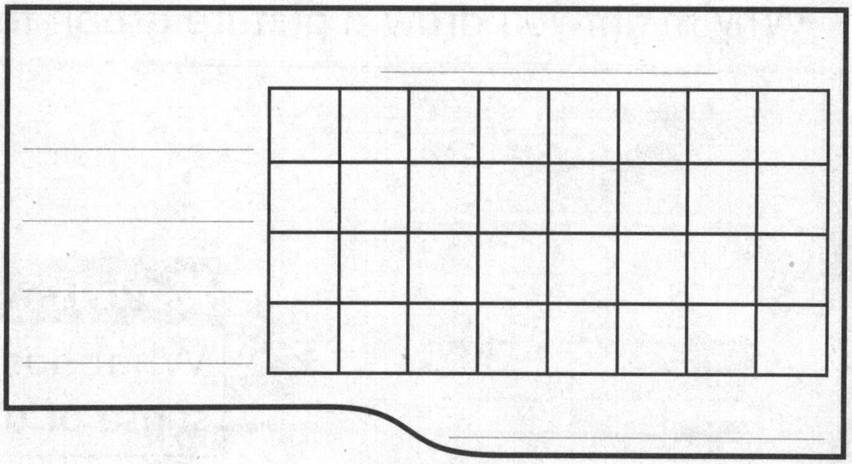

Unit 11 · Data Analysis   173

Use the picture graph to answer the questions.

3. What popcorn flavor was chosen the most?

4. How many people chose butter flavor?

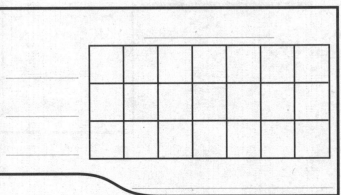

5. **Error Analysis** Kayla says the least favorite flavor is strawberry because it is in the bottom row. How do you respond to Kayla?

6. **Extend Your Thinking** There are 3 yellow houses and 1 blue house on Gio's block. There are 2 more white houses than yellow houses. How can you show the data using a picture graph?

Reflect

Why might you draw a picture graph to show data?

Math is... Mindset

What has helped you make sense of a situation?

174 Lesson 1 • Understand Picture Graphs

Lesson 11-3
Solve Problems Using Bar Graphs

Be Curious

What do you notice?
What do you wonder?

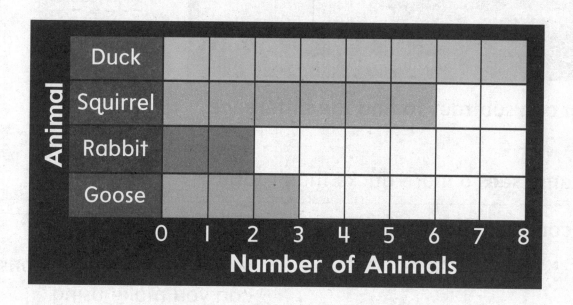

Math is... Mindset

How can deep breaths help you work better?

Unit 11 • Data Analysis 179

Learn

Adama records the animals he sees at the park.

How many more ducks does he see than rabbits?

You can make a bar graph to show the data.

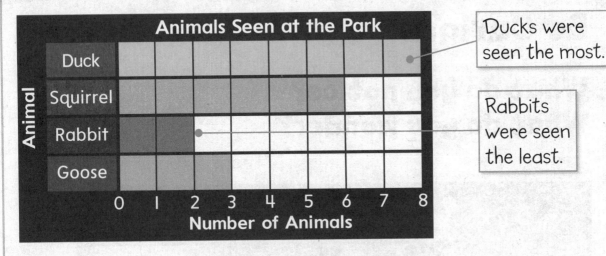

Ducks were seen the most.

Rabbits were seen the least.

You can subtract to find the difference.

8 − 2 = **6**

Adama sees 6 more ducks than rabbits.

You can use a bar graph to solve problems about data.

Math is... Thinking

What other comparisons can you make using this data?

Work Together

How many fewer votes for airplanes are there than cars?

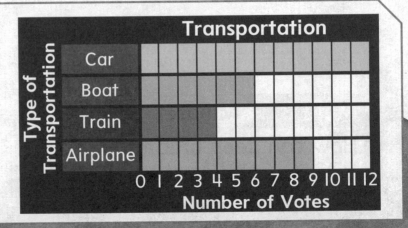

180 Lesson 3 • Solve Problems Using Bar Graphs

On My Own

Name _____

Use the bar graph to answer the questions.

Sienna made a bar graph of her classmates' favorite breakfast foods.

1. **STEM Connection** What is the most popular food?

2. How many more students chose cereal than fruit?

3. How many fewer students chose yogurt than fruit?

4. How many students did not choose the most popular food? Explain your thinking.

Use the bar graph to answer the questions.

Favorite Lunch Food

Food Choice	Number of Students
Sandwich	6
Taco	6
Salad	4
Soup	3

5. How many students chose the 2 most popular lunch foods? Explain your thinking.

6. How many fewer students chose soup than salad?

7. **Extend Your Thinking** Write two questions about the data in the Favorite Lunch Food bar graph. Then answer your questions.

Reflect

Why might you use a bar graph to solve problems?

Math is... Mindset
How have deep breaths helped you work better?

182 Lesson 3 • Solve Problems Using Bar Graphs

Lesson 11-4
Collect Measurement Data

Be Curious

What do you notice?
What do you wonder?

6 inches

4 inches

5 inches

6 inches

5 inches

7 inches

4 inches

5 inches

Math is... Mindset

What do you do to be an active listener?

Learn

Some students measure the lengths of their pencils in inches.

6 inches	5 inches
4 inches	7 inches
5 inches	4 inches
6 inches	5 inches

How can you organize the measurements?

You can make a **tally chart**. A tally chart has columns.

Length of Pencils

Length (inches)	Tally
4	
5	
6	
7	

A tally chart has tally marks. There is one tally mark for each measurement.

Length of Pencils

Length (inches)	Tally
4	II
5	III
6	II
7	I

tally mark

Math is... Precision

Why should you write down your numbers before creating a chart?

You can collect measurement data by measuring the lengths of objects.

 Work Together

Measure 8 classroom objects to the nearest centimeter. Collect the data in a list. Then make a tally chart to show the data.

184 Lesson 4 • Collect Measurement Data

On My Own

Name _____

How can you make a tally chart to show the data?

1. Martin measured the lengths of some pencils.

   5 inches
   4 inches
   5 inches
   5 inches
   5 inches
   7 inches
   7 inches
   5 inches

Length of Pencil	
Length (inches)	Tally

2. Alek measured the lengths of some shoes.

   22 centimeters
   24 centimeters
   25 centimeters
   23 centimeters
   21 centimeters
   24 centimeters
   23 centimeters
   24 centimeters

Length of Shoe	
Length (centimeters)	Tally

3. Measure the lengths of 8 books to the nearest inch. Collect the data in a list. Then make a tally chart to show the data.

Use the data to answer the questions.

4. **Error Analysis** Keya makes a tally chart to show her measurement data. She says her tally chart will have 10 rows. How do you respond to Keya?

| 15 centimeters |
| 16 centimeters |
| 16 centimeters |
| 18 centimeters |
| 19 centimeters |
| 16 centimeters |
| 17 centimeters |
| 16 centimeters |
| 16 centimeters |
| 17 centimeters |

5. How many tally marks go in the row for 19 centimeters?

6. **Extend Your Thinking** How might Keya's tally chart change if she measures 3 more objects that have lengths of 12 centimeters, 14 centimeters, and 20 centimeters?

Reflect

Why is it helpful to organize data in a tally chart?

Math is... Mindset

What did you do to be an active listener?

Lesson 4 • Collect Measurement Data

Lesson 11-5
Understand Line Plots

Be Curious

Tell me everything you can.

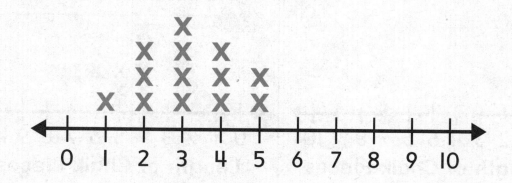

Math is... Mindset

How well do you think you will understand today's lesson?

Learn

Some students measure the lengths of some pieces of chalk. The tally chart shows their measurements.

How can you show the measurements?

Chalk Lengths	
Length (inches)	Tally
2	ǀ
3	
4	ǀǀǀǀ
5	ǀǀ
6	ǀǀǀ

You can make a **line plot**. A line plot is a number line.

Each measurement is one X in the line plot.

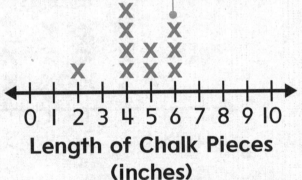

3 pieces are 6 inches long.

Length of Chalk Pieces (inches)

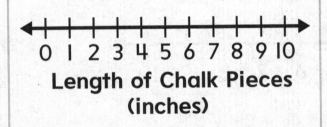

Length of Chalk Pieces (inches)

A line plot is a graph that uses Xs above a number line to display the data.

Math is... Sharing

What other observations can you make about the data?

Work Together

Lana's class measured the length of each student's right foot.

a. How many measurements were recorded?

b. What is the most common length measured?

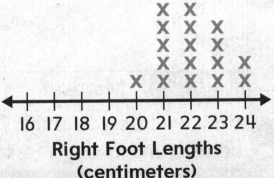

Right Foot Lengths (centimeters)

188 Lesson 5 • Understand Line Plots

On My Own

Name _____

Miss Hart's class measured the lengths of school posters. Use the data in the line plot to answer the questions.

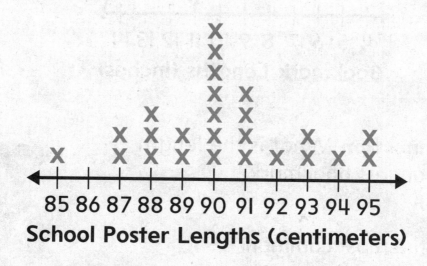

School Poster Lengths (centimeters)

1. What is the most common length measured?

2. What is the least common length measured?

3. What is the length of the longest poster?

4. What is the length of the shortest poster?

5. How many measurements were recorded?

C.J. measured the lengths of his bookmarks. Use the data in the line plot to answer the questions.

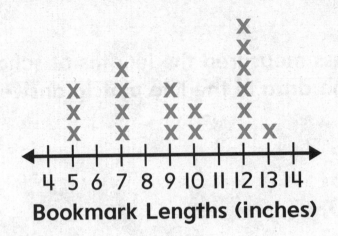

6. **STEM Connection** What is the length of C.J.'s longest bookmark?

7. What is the most common length measured?

8. **Extend Your Thinking** How can you find the difference in length between the longest and the shortest bookmark?

Reflect

Why might you use a line plot to represent data?

Math is... Mindset
How well do you think you understood today's lesson?

Unit II
Reading Line Plots

Name _____

Mr. Shah's class planted a garden. One day, the class measured the height of each plant in the garden. The line plot shows the height of each plant to the nearest inch.

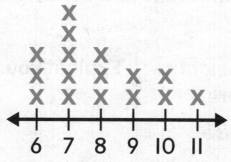

Heights of Plants (inches)

Circle *True* if the statement is true. Circle *False* if it is false.

1. The class measured 6 plants.

   Circle *True* or *False*.

   True   False

Explain your choice.

Unit II · Data Analysis   191

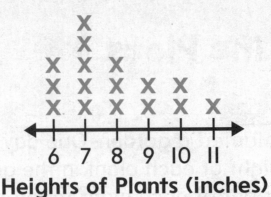

Heights of Plants (inches)

Circle *True* if the statement is true. Circle *False* if it is false.

2. 5 plants have a height of 9 inches or more.

   Circle *True* or *False*.

   True   False

   Explain your choice.

3. The height of the tallest plant is 7 inches.

   Circle *True* or *False*.

   True   False

   Explain your choice.

Reflect On Your Learning

192   Math Probe • Reading Line Plots

Lesson 11-6

Show Data on a Line Plot

Be Curious

How are they the same?
How are they different?

Length (centimeters)	Number of Books
20	l
21	ll
22	
23	ll
24	lll

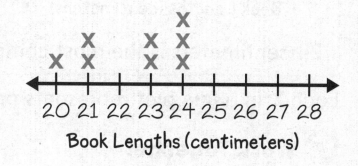

Book Lengths (centimeters)

Math is... Mindset
How can different ideas help you learn better?

Unit 11 · Data Analysis

Learn

Zeke measures the lengths of some of his books.

What length was the most common?

You can show the measurements in a line plot.

Length (centimeters)	Number of Books
20	I
21	II
22	
23	II
24	III

There are 2 Xs above 21 and 23.

There are 3 Xs above 24.

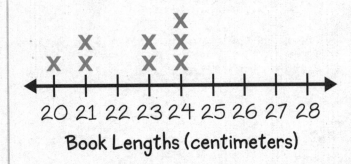

Book Lengths (centimeters)

Math is... Thinking

Why might it be helpful to organize data in a line plot instead of a list?

24 centimeters is the most common length.

Each X in a line plot represents one value in a set of data.

Work Together

How can you represent the measurements using a line plot? Draw a line plot.

Length of Hair

Length (inches)	Students
3	5
4	7
5	3
7	6
9	1

194 Lesson 6 • Show Data on a Line Plot

On My Own

Name _____

How can you represent the measurements using a line plot? Use the data to make a line plot.

1. Samantha measured the heights of toys.

   17 centimeters
   15 centimeters
   10 centimeters
   15 centimeters
   12 centimeters
   17 centimeters
   10 centimeters
   15 centimeters

   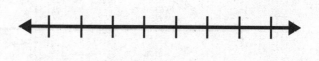

2. John measured the lengths of fish.

   7 inches   5 inches
   12 inches   9 inches
   10 inches   10 inches
   6 inches   12 inches

   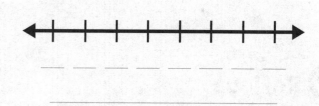

3. Oliver measured the lengths of ties.

   59 inches   59 inches
   58 inches   55 inches
   53 inches   59 inches
   57 inches   58 inches

Unit 11 · Data Analysis   195

How can you use your own measurements to make a line plot? Measure the lengths of 10 used crayons.

4. Record the measurements.

5. Make a line plot of the data.

6. **Extend Your Thinking** Write two questions that can be answered by looking at the line plot.

Reflect

How does a line plot help you show measurements?

Math is... Mindset

How did different ideas help you learn better?

Unit Review   Name

Vocabulary Review

Draw a line to match.

1. tally chart (Lesson 11-1)

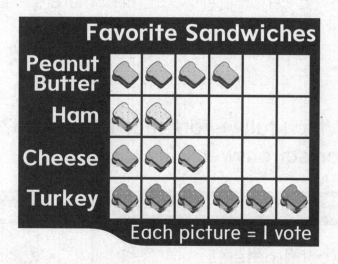

2. line plot (Lesson 11-5)

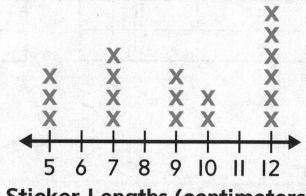

3. picture graph

    (Lesson 11-1)

Unit 11 · Data Analysis   197

Review

4. Jack measures the lengths of some strawberries.

5 centimeters   3 centimeters
4 centimeters   5 centimeters
5 centimeters   6 centimeters
6 centimeters   4 centimeters
3 centimeters   3 centimeters
4 centimeters   5 centimeters

Which tally chart shows the data? Choose the correct answer. (Lesson 11-4)

A. **Length of Strawberry**

Length (centimeters)	Tally
3				
4				
5				
6				

B. **Length of Strawberry**

Length (centimeters)	Tally
3				
4				
5					
6			

C. **Length of Strawberry**

Length (centimeters)	Tally
3				
4				
5			
6					

D. **Length of Strawberry**

Length (centimeters)	Tally
3				
4			
5					
6				

5. Which statement is true about the animals at the petting zoo? Choose all the correct answers. (Lesson 11-3)

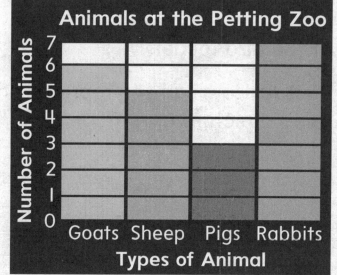

A. There are 9 goats and pigs in all.

B. There are more sheep than goats.

C. There are more pigs and sheep combined than goats.

D. There are 5 fewer rabbits than goats.

6. Damien measures the lengths of some fish and records the data in a line plot. Fill in the blanks. (Lesson 11-5)

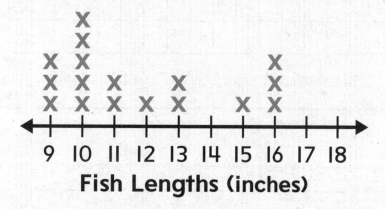

The longest fish are _____ inches long.

The most common length is _____ inches.

Performance Task

Hugo makes a tally chart to show the number of each kind of bird he sees at the bird feeder.

Birds at Bird Feeder	
Bird	Tally
Blue Jay	||||| |||
Cardinal	||||
Robin	||||| |||||
Sparrow	||||| ||||| ||

Part A: How can you represent the data using a picture graph?

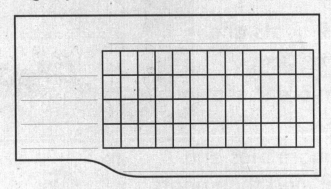

Part B: How can you represent the data using a bar graph?

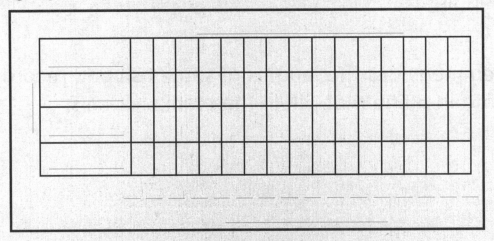

 **Reflect**

How do picture graphs, bar graphs, and line plots help you interpret data?

200 Unit 11 • Performance Task

Unit 11

Fluency Practice

Name _____

Fluency Strategy

> You can use different strategies to add 2-digit numbers. One strategy is to decompose one addend.
>
> $29 + 22 = ?$
>
> $1 + 20 + 1$
>
> **Make a ten:** $29 + 1 = 30$
> **Add tens:** $30 + 20 = 50$
> **Count on:** $50 + 1 = 51$
>
> So, $29 + 22 = $ **51**.

1. What strategies can you use to add $37 + 24$? Show your work.

Fluency Flash

2. What is the sum? Fill in the blanks.

   $58 + 35 = ?$

   ___ + 30 + ___

   Make a ten: $58 + $ _____ $= $ _____
   Add tens: _____ $+ 30 = $ _____
   Count on: _____ $+$ _____ $= $ _____
   So, $58 + 35 = $ _____.

Unit 11 • Data Analysis 201

Fluency Check

What is the sum or difference?

3. 54 − 30 = _____

4. 63 − 7 = _____

5. 39 + 45 = _____

6. 54 + 8 = _____

7. 77 + 5 = _____

8. 43 − 8 = _____

9. 27 + 48 = _____

10. 35 + 27 = _____

11. 65 + 20 = _____

12. 98 − 60 = _____

13. 41 + 35 = _____

14. 46 + 25 = _____

Fluency Talk

What strategies can you use to add 52 + 16? Explain.

How can you make a ten to add 67 + 8? Explain.

Unit 12

Geometric Shapes and Equal Shares

Focus Question

How can I name, draw, and partition geometric shapes?

Hi, I'm Chloe.

I want to be a carpenter. When making steps, I can make rectangles of equal size from one big rectangular board. Knowing about shapes and equal shares will make my job easier.

STEM video | GO ONLINE

Name _____

Prove Me Wrong!

Listen for directions. Use pattern blocks to completely fill these triangles.

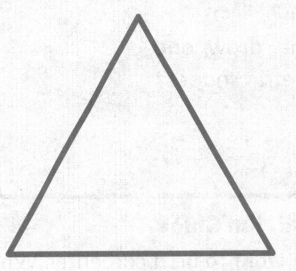

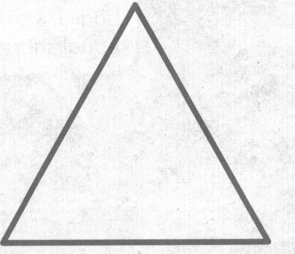

204   Ignite! • Prove Me Wrong!

Lesson 12-2
Draw 2-Dimensional Shapes from Their Attributes

Be Curious

Tell me everything you can.

Math is... Mindset

What helps you make good decisions about your behavior?

Unit 12 • Geometric Shapes and Equal Shares

Learn

How can you draw a 2-dimensional shape given its attributes?

- 3 sides
- 3 angles

- 4 sides
- 4 angles
- all sides the same length

- 4 sides
- 4 angles
- opposite sides the same length

Math is... Exploring

What is the difference between a rectangle and a square?

- 5 sides
- 5 angles
- all sides the same length

- 6 sides
- 6 angles
- all sides different lengths

2-dimensional shapes can be drawn based on their defining attributes.

Work Together

What shape has 5 sides, 5 angles, and all sides different lengths? Draw a shape that matches the attributes. Then write the name.

On My Own

Name _____

Draw a shape that matches the attributes. Then write the name.

1. What shape has 3 sides, 3 angles, and all sides the same length?

2. What shape has 6 sides, 6 angles, and all sides the same length?

3. What shape has 4 sides, 4 angles, and all sides different lengths?

4. **Error Analysis** Maggie says she drew a square. How do you respond to her?

What are 3 defining attributes of the shape?

5.

6.

7.

8. **Extend Your Thinking** Stephen outlined an area of his yard for a garden. The outline has 4 sides and 4 vertices. What shape could the outline be? Explain your thinking and draw 2 possible examples.

Reflect

How does knowing different attributes help you draw 2-dimensional shapes?

Math is... Mindset

What helped you make good decisions about your behavior?

Lesson 12-3
Recognize 3-Dimensional Shapes by Their Attributes

Be Curious

Which doesn't belong?

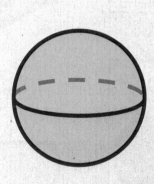

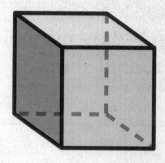
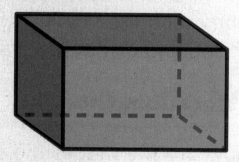

Math is... Mindset

What are some ways to build positive relationships with classmates?

Unit 12 · Geometric Shapes and Equal Shares

Learn

How are the shapes the same? How are they different?

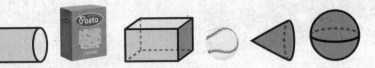

The number of faces, edges, vertices, bases, or having an apex can help you identify 3-dimensional shapes.

	Cube	Rectangular Prism	Sphere
Faces	6 squares	6 rectangles	0
Edges	12	12	0
Vertices	8	8	0
Example			

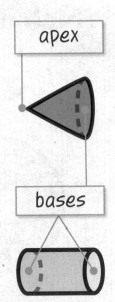

3-dimensional shapes can be recognized by their defining attributes.

Math is... Thinking

What is the difference between 2-dimensional and 3-dimensional shapes?

 Work Together

How are the shapes the same?
How are they different? Explain.

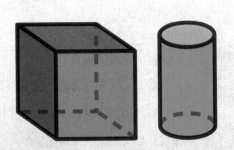

214 Lesson 3 • Recognize 3-Dimensional Shapes by Their Attributes

On My Own

Name _____

How many of each attribute does the shape have? What is the shape?

1.

   ___ faces
   ___ edges
   ___ vertices
   This shape is a _____.

2.

   ___ faces
   ___ edges
   ___ vertices
   This shape is a _____.

3.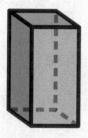

   ___ faces
   ___ edges
   ___ vertices
   This shape is a _____.

4.

   ___ base
   ___ apex
   This shape is a _____.

5. Which shapes are rectangular prisms? Choose all the correct answers.

   A.    B.   C.    D.

Unit 12 · Geometric Shapes and Equal Shares

6. Which shapes are spheres? Choose all the correct answers.

A.  B.   C.  D.

7. **STEM Connection** Sienna is serving water to runners at a marathon. What shape are the cups? Explain.

8. **Extend Your Thinking** Aisha has an object with 6 faces, 12 edges, and 8 vertices. What shape could the object be? Explain.

 Reflect

How can you identify 3-dimensional shapes?

Math is... Mindset

How did you build positive relationships with classmates?

Lesson 12-4
Understand Equal Shares

Be Curious

**How are they the same?
How are they different?**

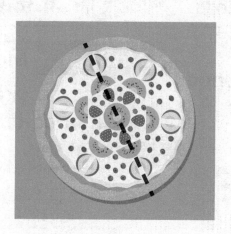

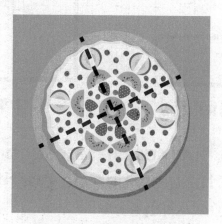

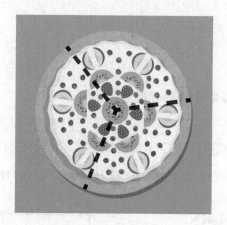

Math is... Mindset

How can your strengths help you learn today?

Unit 12 • Geometric Shapes and Equal Shares

Learn

Some friends are using this paper to make crafts.

What are some different ways they can share each paper between either 2, 3, or 4 friends?

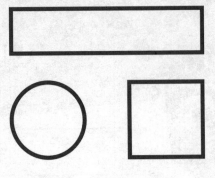

Shares that are the same size are **equal shares**.

2 equal shares   3 equal shares   4 equal shares

2 halves   **3 thirds**   **4 fourths**

Shapes, such as circles, squares, and rectangles, can be **partitioned** into equal shares.

Math is... Modeling

How can a circle be partitioned into 3 equal shares?

Work Together

How can you partition the rectangle into 4 equal shares? Draw to show your work.

218 Lesson 4 • Understand Equal Shares

On My Own

Name _____

Which shapes are partitioned into equal shares? Choose all the correct answers.

1. A.    B.    C.

2. A. 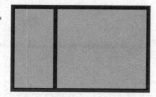   B. 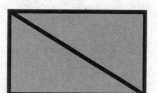   C.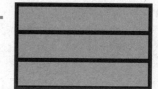

3. A.    B.    C.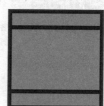

4. A. 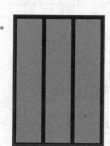   B. 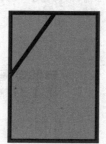   C.

5. How can you partition the circle into 2 equal shares? Draw to show your work.

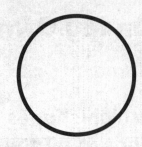

6. How can you partition the square into 3 equal shares? Draw to show your work.

7. How can you partition the rectangle into 4 equal shares? Draw to show your work.

8. **Extend Your Thinking** How can you partition a shape that has 4 sides, 4 angles, and all sides the same length into 2 equal shares? Draw to show your work. Aubree thinks the shape will be a rectangle. How do you respond to her?

Reflect

How can you partition rectangles, circles, and squares into equal shares?

Math is... Mindset
How have your strengths helped you learn today?

Unit 12
Partitioning Shapes

Name _____

Decide if each shape has been partitioned into four equal shares. Circle *Yes* or *No*.

1.

   Are there four equal shares?

   Yes   No

   Explain why you chose Yes or No.

2.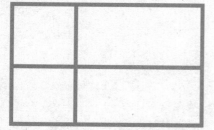

   Are there four equal shares?

   Yes   No

   Explain why you chose Yes or No.

Unit 12 • Geometric Shapes and Equal Shares

Decide if each shape has been partitioned into four equal shares. Circle Yes or No.

3.

   Are there four equal shares?

   Yes   No

   Explain why you chose Yes or No.

4.

   Are there four equal shares?

   Yes   No

   Explain why you chose Yes or No.

Reflect On Your Learning

Lesson 12-5
Relate Equal Shares

Be Curious

Tell me everything you can.

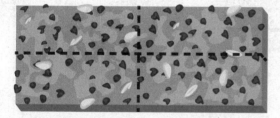
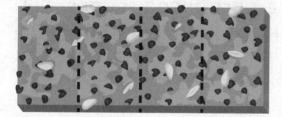

Math is... Mindset

What helps you stay focused on your work?

Learn

Olive says you can partition these shapes into 2, 3, or 4 equal shares in different ways.

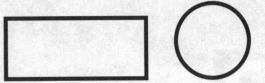

How can you relate the equal shares?

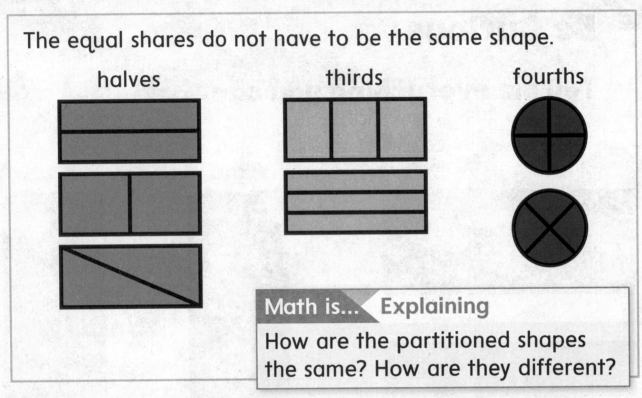

Math is... Explaining

How are the partitioned shapes the same? How are they different?

Shapes, such as circles or rectangles, can be partitioned into equal shares in different ways.

Work Together

How can you partition the square into fourths? Show three different ways.

224 Lesson 5 • Relate Equal Shares

On My Own

Name _____

Choose all the correct answers.

1. Which shows a circle partitioned into halves?

   A.    B.    C.

2. Which shows a rectangle partitioned into thirds?

   A.    B.    C.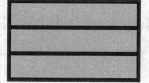

3. Which shows a square partitioned into fourths?

   A. 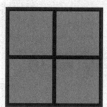   B. 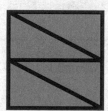   C.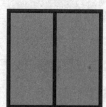

4. How can you partition the circle into equal shares? Show two different ways.

Unit 12 • Geometric Shapes and Equal Shares

5. **Error Analysis** Selena partitions a rectangle into thirds. Brian partitions the same rectangle into thirds. Their shares are different shapes. Selena and Brian think their shares are not equal because they are not the same shape. How would you respond to them?

6. **Extend Your Thinking** A slice of cinnamon bread is in the shape of a square. Draw a picture to explain how to partition the slice of bread to split it equally between 4 friends. How much of the slice of bread does each friend get?

Reflect

Why can shapes be partitioned into equal shares in more than one way?

Math is... Mindset

What has helped you stay focused on your work?

Lesson 12-6

Partition a Rectangle into Rows and Columns

Be Curious

**What do you notice?
What do you wonder?**

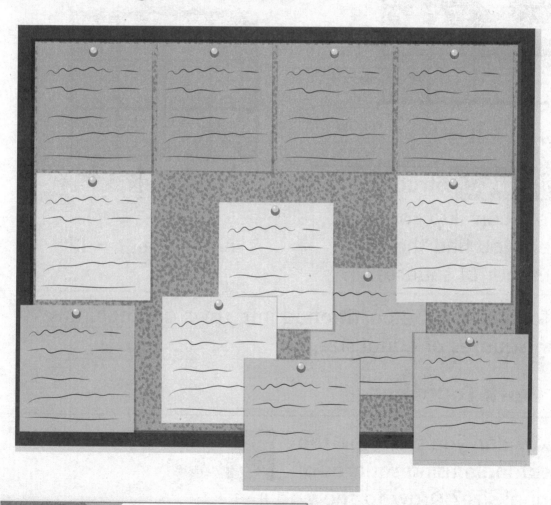

Math is... Mindset

What helps you understand your classmates' ideas?

Unit 12 · Geometric Shapes and Equal Shares

Learn

How can you find the number of squares that will fill the rectangle?

You can use repeated addition to find the number of squares.

▶ **One Way** Add the rows. Each row has 4 squares.

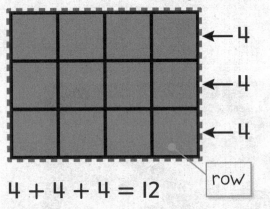

$4 + 4 + 4 = 12$

> **Math is... Structure**
>
> How can skip counting help you find the total number of squares?

▶ **Another Way** Add the columns. Each column has 3 squares.

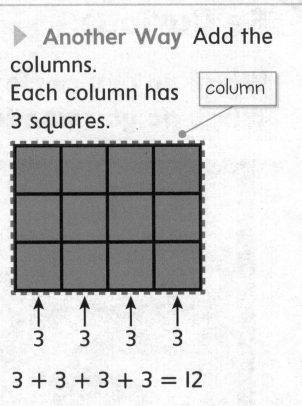

$3 + 3 + 3 + 3 = 12$

Rectangles can be partitioned into rows and columns using squares of equal size.

Work Together

How can you partition the rectangle using squares of equal size? Draw to show your work. How many squares can you partition the rectangle into?

Total squares:

228 Lesson 6 • Partition a Rectangle into Rows and Columns

On My Own

Name _____

How many rows, columns, and squares is the rectangle partitioned into? Write an equation to find the total number of squares.

1.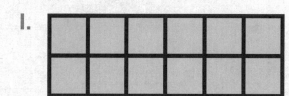
   a. Rows: ____
   b. Columns: ____
   c. Equation: _____
   d. Total squares: ____

2.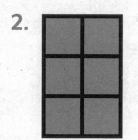
   a. Rows: ____
   b. Columns: ____
   c. Equation: _____
   d. Total squares: ____

3.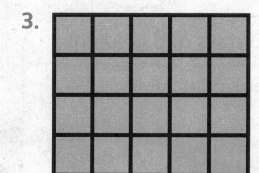
   a. Rows: ____
   b. Columns: ____
   c. Equation: _____
   d. Total squares: ____

How can you partition the rectangle using squares of equal size? Draw to show your work. What is the total number of squares?

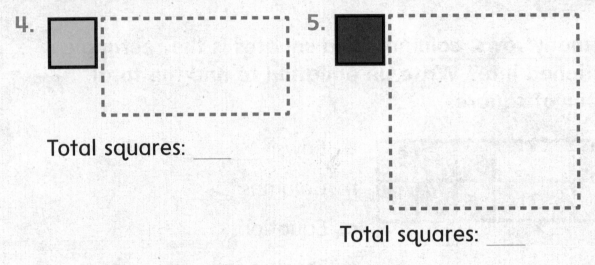

4.

Total squares: ____

5.

Total squares: ____

6. **Extend Your Thinking** Leo and his sister want to partition their rectangular garden into square plots. Leo says there can be 3 square plots. His sister says there can be 12 square plots. Who do you agree with? Draw a picture to show why.

Reflect

How can you partition a rectangle into rows and columns using squares of equal size?

Math is... Mindset

What has helped you understand your classmates' ideas?

Unit Review   Name

Vocabulary Review

Draw a line to match.

1. **fourths**
   (Lesson 12-4)

2. **halves**
   (Lesson 12-4)

3. **pentagon**
   (Lesson 12-1)

4. **quadrilateral**
   (Lesson 12-1)

5. **thirds**
   (Lesson 12-4)

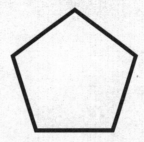

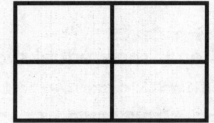

Unit 12 • Geometric Shapes and Equal Shares   **231**

Review

6. Which shapes are spheres? Choose all the correct answers. (Lesson 12-3)

   A.

   B.

   C.

   D.

7. Which shapes show equal shares? Choose all the correct answers. (Lesson 12-4)

   A.   B.   C.    D.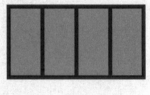

8. Which shapes have 5 sides, 5 angles, and 5 vertices? Choose all the correct answers. (Lesson 12-1)

   A.

   B.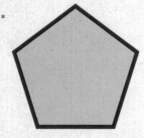

   C.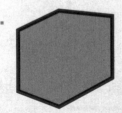

   D.

9. Mr. Johnson partitions a gym floor that is shaped like a rectangle. Show two ways he could partition the gym floor into halves. Draw lines to show your work.

(Lesson 12-5)

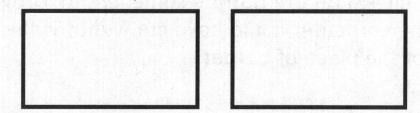

10. Nina drew a shape that has 3 sides and 3 angles, where all of the sides are the same length. Which shape did Nina draw? (Lesson 12-2)

A. 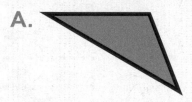   B.

C. 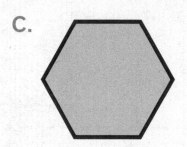   D.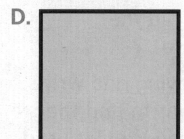

11. How can you partition the rectangle using squares of equal size? Draw lines to show your work. (Lesson 12-6)

Unit 12 • Geometric Shapes and Equal Shares

Performance Task

A carpenter remodeled a bedroom and bathroom in his house.

Part A: A carpenter cut a piece of carpet for a bedroom. It has 4 angles and 4 sides. The opposite sides are the same length, but all 4 sides are not the same length. Draw a piece of carpet the carpenter could have cut. What is the name of the shape of the piece of carpet?

Part B: A carpenter used square tiles for the back wall in a shower. How many square tiles did the carpenter use?

Make a drawing and write two equations to find the number of square tiles used.

 Reflect

How can you name, draw, and partition geometric shapes?

Unit 12
Fluency Practice

Name _____

Fluency Strategy

> You can use many strategies to subtract 2-digit numbers. One way is to decompose one number in the equation.
>
> 58 − 43 = ?   **Subtract tens:** 58 − 40 = 18
>   ╱╲   **Count back:** 18 − 3 = 15
> 40 + 3
>
> So, 58 − 43 = 15.

1. What strategies can you use to subtract 72 − 38? Show your work.

Fluency Flash

2. What is the difference? Fill in the blanks.

   44 − 21 = ?   Subtract tens: 44 − _____ = _____
     ╱╲   Count back: _____ − _____ = _____
    ___ + ___   So, 44 − 21 = _____.

Fluency Check

What is the sum or difference?

3. 37 − 19 = _____

4. 64 + 19 = _____

5. 71 + 26 = _____

6. 52 − 4 = _____

7. 82 − 49 = _____

8. 45 + 13 = _____

9. 65 − 8 = _____

10. 77 − 24 = _____

11. 64 − 23 = _____

12. 45 − 31 = _____

13. 28 + 32 = _____

14. 67 − 49 = _____

Fluency Talk

What strategies can you use to subtract 53 − 36? Explain your thinking.

What strategies can you use to add 15 + 76? Explain.

Glossary/Glosario

English	Spanish/Español
Aa	
a.m. The hours from midnight until noon.	**a.m.** Las horas que van desde la medianoche hasta el mediodía.
add (adding, addition) To join together sets to find the total or sum. 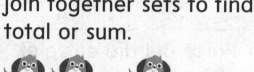 4 + 3 = 7	**sumar (adición)** Unir conjuntos para hallar el total o la suma.  4 + 3 = 7
addend Any numbers or quantities being added together.  2 is an addend and 3 is an addend	**sumando** Cualquieres números o cantidades que se suman. 2 + 3 2 es un sumando y 3 es un sumando

Glossary GI

English	Spanish/Español
adjusting For addition, take some from one number and give to another number to make the problem easier to solve. For subtraction, take the same amount from both numbers or give the same amount to both numbers to make the problem easier to solve.	**ajuste** Tomar de un número y dárselo a otro número para que el problema sea más fácil de resolver.
afternoon The part of the day between noon and sunset.	**tarde** Parte del día entre el mediodía y la puesta del sol.
array Objects displayed in rows and columns. 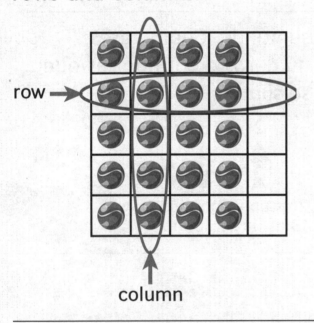	**arreglo** Objetos presentados en filas y columnas. 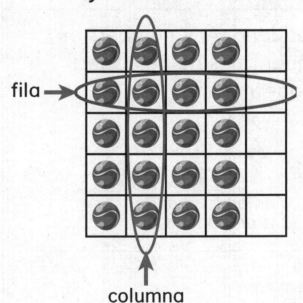

English | Spanish/Español

Bb

bar graph A graph that uses bars to show data.

gráfica de barras Gráfica que usa barras para ilustrar datos.

Cc

cent

centavo

1 cent   1 ¢

1 centavo   1 ¢

cent sign (¢) The sign used to show cents.

centavo (¢) El signo que se usa para mostrar centavos.

1 ¢   5 ¢

1 ¢   5 ¢

centimeter A metric unit for measuring length.

centímetro Unidad métrica para medir la longitud.

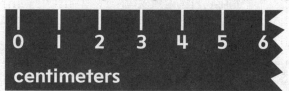

Glossary G3

English	Spanish/Español
circle A closed, round figure. 	**círculo** Figura redonda y cerrarda.
column A column goes up and down on a number chart.	**columna** Una columna sube y baja en una tabla numérica.
compare To look at objects, shapes, or numbers and see how they are alike or different.	**comparar** Observar objetos, formas o números para saber en qué se parecen y en qué se diferencian.
count back On a number line, start at the greater number and count back. 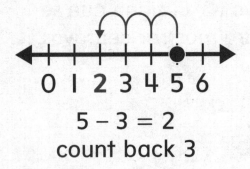 $5 - 3 = 2$ count back 3	**contar hacia atrás** En una fila de números, comienza en el número 5 y cuenta 3 hacia atrás. 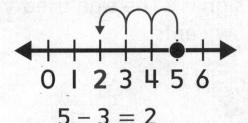 $5 - 3 = 2$ cuenta 3 hacia atrás

English	Spanish/Español
count on Start at a number on a number line and count up to another number. 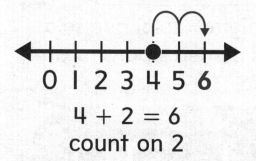 $4 + 2 = 6$ count on 2	**contar hacia adelante** Comenzar en un número en una recta numérica y contar hasta el siguiente número.  $4 + 2 = 6$ cuenta 2 hacia adelante

Dd

data Numbers or symbols collected to show information.

Name	Number of Pets
Mary	3
James	1
Alonzo	4

datos Números o símbolos que se reúnen para mostrar información.

Nombre	Número de mascotas
Mary	3
James	1
Alonzo	4

decompose To break a number into different parts.

descomponer Separar un numero de diferentes partes.

difference The answer to a subtraction problem.

$3 - 1 = 2$

The difference is 2.

diferencia Respuesta a un proble ma de resta.

$3 - 1 = 2$

La diferencia es 2.

Glossary G5

English	Spanish/Español
digit A symbol used to write numbers. The ten digits are: 0, 1, 2, 3, 4, 5, 6, 7, 8, 9	**dígito** Símbolo usado para escribir números. Los diez dígitos son: 0, 1, 2, 3, 4, 5, 6, 7, 8, 9
digital clock A clock that uses only numbers to show time. 	**reloj digital** Reloj que sólo utiliza números para mostrar la hora. 
dime dime = 10¢ or 10 cents head tail	**dime** moneda de 10¢ = 10¢ o 10 centavos cara cruz
dollar One dollar = 100¢ or 100 cents. Also written as $1 or $1.00.   front back	**dólar** Un dólar = 100¢ o 100 centavos. También se escribe como $1 o $1.00.  frente revés
dollar sign ($) The sign used to show dollars. one dollar = $1 or $1.00	**signo de dólar ($)** Símbolo que se usa para mostrar dólares. un dólar = $1 o $1.00

English	Spanish/Español
doubles Two addends that are the same number. $6 + 6 = 12$	**dobles** Dos sumandos que son el mismo número. $6 + 6 = 12$

Ee

equal groups Each group has the same number of objects. ●●●● ●●●● ●●●● ●●●● There are two equal groups of counters.	**grupos iguales** Cada grupo tiene el mismo número de objetos. ●●●● ●●●● ●●●● ●●●● Hay dos grupos iguales de fichas.
equal shares Each share is the same size. Example: This sandwich is cut into 2 equal shares. 	**partes iguales** Cada una de las partes tiene el mismo tamaño. Ejemplo: Este pastelillo está cortado en 2 partes iguales. 

English	Spanish/Español
equal to (=) 🏈🏈🏈 🏈🏈🏈 🏈🏈🏈 🏈🏈🏈 6 = 6 6 is equal to or the same as 6	equal a (=) 🏈🏈🏈 🏈🏈🏈 🏈🏈🏈 🏈🏈🏈 6 = 6 6 es igual o lo mismo que 6
estimate To find a number close to an exact amount. 107 is close to 100.	estimado Hallar un número cercano a la cantidad exacta. 107 es cercano a 100.
even number Any number with 0, 2, 4, 6, or 8 in the ones place.	número par Los números que terminan en 0, 2, 4, 6, 8.
expanded form The representation of a number as a sum that shows the value of each digit. 536 is written as 500 + 30 + 6.	forma desarrollada Representación de un número como una suma que muestra el valor de cada dígito. 536 se escribe como 500 + 30 + 6.

Ff

English	Spanish/Español
foot A unit to measure length. The plural is feet.  12 inches = 1 foot	pie Una unidad para medir longitud.  12 pulgadas = 1 pie
fourths Four equal parts of a whole. Each part is a fourth, or a quarter of the whole.	cuartos Cuatro partes iguales de un todo. Cada parte es un cuarto, o la cuarta parte del todo.

English | Spanish/Español

Gg

greater than (>)

7 > 2

7 is greater than 2.

mayor que (>)

7 > 2

7 es mayor que 2.

Hh

halves Two equal parts of a whole. Each part is a half of the whole.

mitades Dos partes iguales de un todo. Cada parte es la mitad de un todo.

hexagon A 2-dimensional shape that has 6 sides.

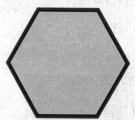

hexágono Una figura bidimensional con 6 lados.

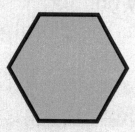

hour A unit of time.
1 hour = 60 minutes

hora Unidad de tiempo.
1 hora = 60 minutos

Glossary G9

English	Spanish/Español
hour hand The hand on a clock that tells the hour. It is the shorter hand.  hour hand	**manecilla horaria** Manecilla del reloj que indica la hora. Es la manecilla más corta.  manecilla horaria
hundreds The numbers 100–999. Example: In the number 234, 2 is in the hundreds place. 234 hundreds place	**centenas** Los números 100–999. Ejemplo: En el número 234, el 2 está en el lugar de las centenas. 234 ↑ lugar de las centenas

Ii

inch A customary unit for measuring length. The plural is inches.  12 inches = 1 foot	**pulgada** Unidad habitual para medir longitud.  12 pulgadas = 1 pie

English | Spanish/Español

Kk

key Tells what or how many each symbol stands for.

clave Nos dice qué o cuánto representa cada símbolo.

Ll

length How long or how far away something is.

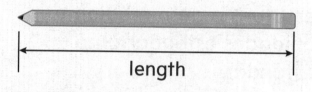

longitud La mayor de las dos dimensiones principales que tienen las cosas o figuras planas.

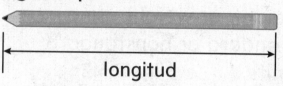

less than (<)

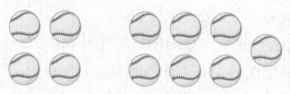

4 < 7

4 is less than 7.

menor que (<)

4 < 7

4 es menor que 7.

English	Spanish/Español
line plot A graph that uses columns of Xs above a number line to show frequency of data.	**diagrama de puntos** Gráfico que usa columnas de X sobre una recta numérica para mostrar la frecuencia de los datos.

Grade in School

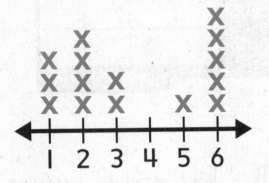

Grado en la escuela

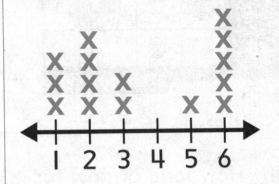

Mm

measure To find the length, height, or weight using standard or nonstandard units.	**medir** Hallar la longitud, estatura o peso mediante unidades estándar o no estándar.
meter A metric unit for measuring length. It is about the length of a baseball bat or the width of a door.	**metro** Unidad métrica para medir longitud. Es aproximadamente del largo de un bate de béisbol o del ancho de una puerta.

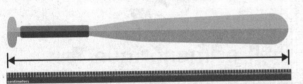

1 meter = 100 centimeters

1 metro = 100 centímetros

English	Spanish/Español
midnight The middle of the night. 12:00 at night	**medianoche** La mitad de la noche. Las 12:00 a.m.
minute A unit used to measure time. 1 minute = 60 seconds	**minuto** Unidad para medir tiempo. 1 minuto = 60 segundos
minute hand The longer hand on a clock that tells the minutes. 	**minutero** La manecilla más larga del reloj. Indica los minutos. 
missing addend In an addition equation, the sum and one addend are known, and the missing addend is unknown. 9 + ? = 16 The missing addend is 7.	**sumando que falta** En una ecuación de suma, se conoce la suma y un sumando y el sumando que falta es desconocido. 9 + ? = 16 El sumando que falta es 7.

English	Spanish/Español
## Nn	
nickel nickel = 5¢ or 5 cents head tail	nickel moneda de 5¢ = 5¢ o 5 centavos cara cruz
noon The middle of the day. 12:00 in the afternoon	mediodía La mitad del día. Las 12 p.m.
number line A line with number labels. 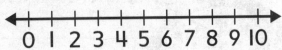	recta numérica Recta con marcas de números. 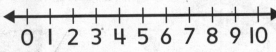
## Oo	
odd number Any number with 1, 3, 5, 7, or 9 in the ones place.	número impar Los números que terminan en 1, 3, 5, 7, 9.
ones The numbers in the range of 0–9. A place value of a number. 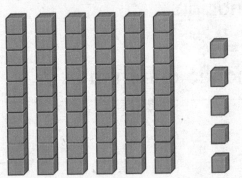 65 5 is in the ones place.	unidades Los números en el rango de 0 a 9. Valor posicional de un número. 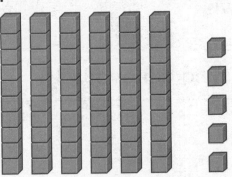 65 El 5 está en el lugar de las unidades.

English	Spanish/Español
Pp	
p.m. The hours from noon until midnight.	**p.m.** Las horas que van desde el mediodía hasta la medianoche.
partial sums A step-by-step process to add one place value at a time, and then add those sums to find the total sum. 42 + 17 Decompose 42 into 40 and 2, and 17 into 10 and 7. Add the tens: 40 + 10 = 50 Add the ones: 2 + 7 = 9 Add the partial sums: 50 + 9 = 59	**sumas parciales** Proceso paso a paso para sumar un lugar posicional a la vez, y luego sumar los resultados para hallar la suma total. 42 + 17 Descomponer 42 en 40 y 2, y 17 en 10 y 7. Sumar las decenas: 40 + 10 = 50 Sumar las unidades: 2 + 7 = 9 Sumar los resultados parciales: 50 + 9 = 59
partition To divide or break up.	**separar** Dividir o desunir.
pattern An order that a set of objects or numbers follows over and over. 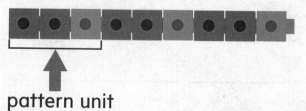 pattern unit	**patrón** Orden que sigue continuamente un conjunto de objetos o números. 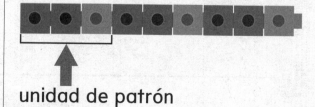 unidad de patrón

English	Spanish/Español
penny penny = 1¢ or 1 cent head tail	penny moneda de 1¢ = 1¢ o 1 centavo cara cruz
pentagon A figure with 5 sides. 	pentágono Un polígono con cinco lados.
picture graph A graph that has different pictures to show data collected. 	gráfica con imágenes Gráfica que tiene diferentes imágenes para ilustrar la información recopilada. 

Qq

quarter quarter = 25¢ or 25 cents head tail	quarter moneda de 25¢ = 25¢ o 25 centavos cara cruz

Rr

rectangle A shape with 4 sides and 4 angles.	rectángulo Figura con 4 lados y 4 esquinas.

English	Spanish/Español
regroup To take apart a number to write it in a new way. 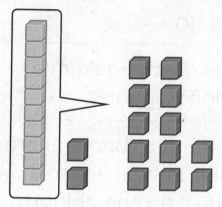 1 ten + 2 ones becomes 12 ones	**reagrupar** Separar un número para escribirlo en una nueva forma. 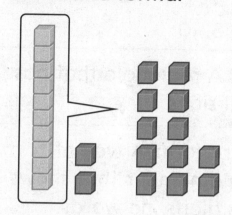 1 decena + 2 unidades se convierten en 12 unidades
row A row goes left to right on a number chart.	**fila** Una fila se lee de izquierda a derecha en una tabla numérica.

Ss

side One of the lines that make up a shape. 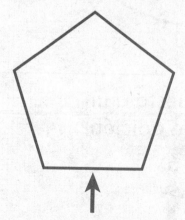 A pentagon has 5 sides.	**lado** Uno de la lãneas que compone una figura. 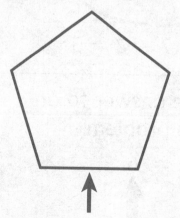 El pentágono tiene cinco lados.

Glossary G17

English	Spanish/Español
skip count To count objects in equal groups of two or more. 2, 4, 6, 8, 10	**contar salteado** Contar objetos en grupos iguales de dos o más. 2, 4, 6, 8, 10
square A rectangle that has 4 equal sides.	**cuadrado** Rectángulo que tiene 4 lados iguales.
standard form A way of writing a number that shows only its digits, no words. 537 89	**forma estándar** Una manera de escribir un número solo con dígitos, no con palabras. 537 89
subtract (subtracting, subtraction) To take away, take apart, separate, or find the difference between two sets. The opposite of addition. 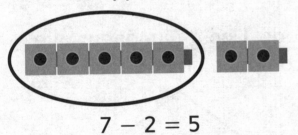 $7 - 2 = 5$	**restar (resta, sustracción)** Eliminar, quitar, separar o hallar la diferencia entre dos conjuntos. Lo opuesto de la suma. 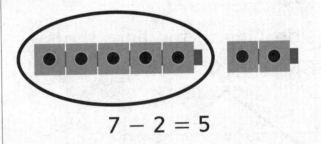 $7 - 2 = 5$

sum The answer to an addition problem.
 	
2 + 4 = **6**
         ↑
        sum
	suma Respuesta a un problema de adición.
	
	2 + 4 = **6**
	         ↑
	        suma

English	Spanish/Español
survey To collect data by asking people the same questions.	**encuesta** Recolectar datos haciendo las mismas preguntas a las personas.

Favorite Color	
Color	Tally
Blue	ⵑⵑⵑⵑ l
Yellow	llll
Red	ⵑⵑⵑⵑ lll

Color Preferido	
Color	Marca
Azul	ⵑⵑⵑⵑ l
Amarillo	llll
Rojo	ⵑⵑⵑⵑ lll

This tally chart shows the results from a survey.

Esta tabla de conteo muestra los resultados de una encuesta.

Tt

tally chart A way to show data collected using tally marks.

tabla de conteo Una manera de mostrar los datos obtenidos usando marcas de conteo.

Favorite Sport	
Sport	Tally
⚽	ⵑⵑⵑⵑ
🏀	ⵑⵑⵑⵑ llll
⚾	ⵑⵑⵑⵑ ll

Deporte preferido	
Deporte	Marca
⚽	ⵑⵑⵑⵑ
🏀	ⵑⵑⵑⵑ llll
⚾	ⵑⵑⵑⵑ ll

tally mark(s) A mark used to record data collected in a survey.
ⵑⵑⵑⵑ ll

marca(s) Símbolo usado para anotar datos de una encuesta.
ⵑⵑⵑⵑ ll

English	Spanish/Español
tens A place value of a number. 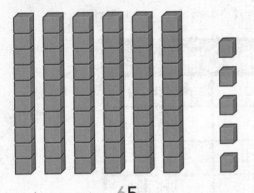 6 is in the tens place.	**decenas** Valor del lugar de un número. 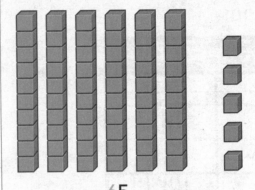 6 está en el lugar de las decenas.
thirds Three equal parts.	**tercios** Tres partes iguales.
trapezoid A four-sided shape with only two opposite sides that are the same length.	**trapecio** Figura de cuatro lados con solo dos lados opuestos que son paralelos. 
triangle A shape with 3 sides and 3 angles. 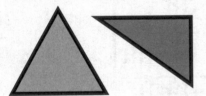	**triángulo** Figura con 3 lados y 3 esquinas. 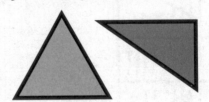

Uu

unit An object used to measure.	**unidad** Objeto que se usa para medir.

English	Spanish/Español
unknown A missing number in an equation. $9 + ? = 10$	**incógnita** El número que falta en una ecuación. $9 + ? = 10$

Ww

whole The entire object. 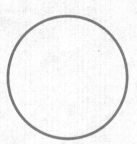	**el todo** El objeto completo. 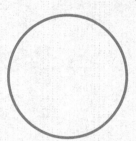
word form A form of a number that uses written words. 472 four hundred seventy-two	**en palabras** Forma de escribir un número en palabras. 472 cuatrocientos setenta y dos

Yy

yard A unit of measure for length. 1 yard = 3 feet  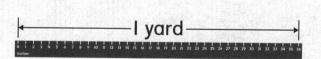	**yarda** Unidad de medida de longitud. 1 yarda = 3 pies  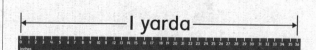

Glossary G21